Student Study Guide
to accompany

ELEMENTARY STATISTICS
A Step by Step Approach

Second Edition
Allan G. Bluman

Prepared by
Pat Foard
South Plains College

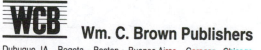
WCB **Wm. C. Brown Publishers**

Dubuque, IA   Bogota   Boston   Buenos Aires   Caracas   Chicago
Guilford, CT   London   Madrid   Mexico City   Sydney   Toronto

Contents

PREFACE

This Study Guide is intended to accompany <u>Elementary Statistics; A Step by Step Approach</u> by Allan G. Bluman. It can also be used with most other elementary statistics texts. The student should use this study guide to reinforce what you have learned in the book or learned in class. The best way for you to use this study guide would be to read the text and listen to the class lecture first. Then read the section called Understanding the Topic and answer the questions to check your understanding of the topic. Then you should follow each example in the application section and immediately see if you can work the accompanying exercise. If you have any problems at this point you should see your instructor for help. The practice test can be used to see if you understand all the concepts for the whole chapter.

This Study Guide includes:

Understanding the Topic - a chapter summary to reinforce the main ideas of the text.

Checking Your Understanding - written questions to allow you to make sure you know the ideas before you start the exercises.

Applying Your Understanding - an example carefully explained and an exercise using the same concepts immediately following.

Practice Test - to tie all the concepts of the chapter together.

I wish to thank Don Foard and Jesse Rincones for help in getting this book ready for publication. I would also like to thank James Stephens for help with the graphs in this publication.

Pat Foard

CHAPTER 1
THE NATURE OF PROBABILITY
AND STATISTICS
Understanding Statistics

INTRODUCTION

Statistics are used by almost everybody in almost every job. Sports, medicine, science, education, and business fields all use statistics in everyday work. There are two branches of statistics: descriptive and inferential.

TYPES OF STATISTICS

Descriptive statistics are all the methods used to collect, organize, or present data, usually to make the data easier to understand. If you present numbers in a graph or average grades to get a better idea how a class is doing, you are using descriptive statistics.

Inferential statistics are the methods used to collect, summarize, or present data to help in decision making. If you use a small group to predict what a whole population will do or if you compare test results from two different classes to decide if one teaching method is better, then you are using inferential statistics.

SPECIAL DEFINITIONS

A **variable** is a characteristic that can have different values; it can "vary." A collection of variables is called **data**. If you wrote down the ages of everyone in a room, the ages would be the variable and all the numbers that you wrote down would be your data. Variables are classified as either qualitative or quantitative. **Qualitative data** can be placed in categories, such as gender (male or female). **Quantitative data** is numerical values. Quantitative data can be ordered and ranked.

Quantitative variables can be classified into one of two groups. **Discrete variables** are values that are obtained by counting. The results will be whole numbers. **Continuous variables** are values that are obtained by measuring. The results can be any value between two specific values. If you count everyone in a room, the variables will be discrete, but if you take everyone's height, you could get any number between two reasonable amounts, so heights are continuous variables.

A **population** is a collection of all the objects or subjects to be studied, but a small group or subset of a population is a **sample**. The population must be carefully defined to know exactly who the whole group is. If you were interested in how all the voters in the United States were going to vote in an election and interviewed all the voters in Earth, Texas, the population would be every registered voter in the United States and the voters in Earth are a sample. But if you were interested in the outcome of a local election and asked every voter in Earth, then you have interviewed the whole population.

LEVELS OF MEASUREMENT

Variables can also be classified according to the level of measurement. There are four levels of measurement: nominal, ordinal, interval and ratio.

Data at the **nominal level of measurement** can be classified into groups, but no order or rank can be established. Often nominal level of data consists of names. Data that consists of sex, religious preferences, or college majors is nominal level.

Data at the **ordinal level of measurement** can be ordered or ranked, but a precise difference in the levels cannot be determined. Letter grades are an example of ordinal level data. You can order grades, A is better than B and so on, but there is not an exact difference between any two letter grades. An A and B might be close (91 and 88) or might be far apart (99 and 80), so the exact differences cannot be determined.

Interval level of measurement is data that can be ordered and has an exact difference between any two units, but has no meaningful zero or starting point. Temperatures are an interval level because they can be ordered, there is an exact difference between any two degrees, but the zero does not mean the starting point since there can be temperatures below zero.

Ratio level of measurement is the highest level of measurement. Data at this level can be ordered, has an exact difference between units, and has a meaningful zero. Things that are counted are usually ratio level.

DATA COLLECTION

Data can be collected in various ways. Surveys, researching records, or direct observations are some of the ways to collect data. The methods of survey that are used the most often are telephone, mail-in, or personal interview.

Sampling techniques are used when part of a population is to be surveyed. If it takes too long or is too expensive to interview the whole population, a sample is used. If a sample is chosen correctly so it represents the population, it is called **unbiased**. If the sample does not represent the whole population, it is **biased**.

Random sampling is used to see that all the possible elements of the population have an equal opportunity of being selected for the sample. One method to obtain a random sample is to number all the elements in a population, mix the numbers thoroughly and draw out how many numbers you want to sample and interview just the ones whose numbers were chosen. Another way to choose the numbers without mixing and drawing is to use a random number chart. Table D in the back of this workbook is a random number chart. Close your eyes and point to the chart to see what number to start with. If you want to sample 10 items, write down 10 numbers from the chart starting with the one you picked and sample those 10 subjects. Notice the table has 5 digits in each number. You read as many of the digits as you need. If you only need two digits, read the last 2 digits from each number.

Systematic sampling numbers the population and then selects numbers at a regular interval. Depending on how many you want to sample, you might take every third or every eighth number. To use systematic sampling, the population must be randomly mixed when you number them.

Stratified sampling divides the population into subgroups that have characteristics that might be important to the study. A population might be divided according to sex, age, or income level if these are important. Then a number from each group that is proportional to the number of each group in the population will be chosen to be studied.

Cluster sampling samples an already existing group. One class might be used as a sample for a whole school. Cluster sampling is often easier and cheaper than other methods, but sometimes the clusters do not represent the whole group.

Checking Your Understanding

Complete this section before you go on. Write in the book. The answers are in the back of the book. Make a note of any questions that you wish to discuss with your instructor.

I. Discuss the differences between inferential and descriptive statistics.

II. Discuss the differences between discrete and continuous variables. Give an example of each that is not given in this book or the text.

III. Discuss the differences between qualitative and quantitative data.

IV. Select the correct answer and write the appropriate letter in the space provided.

_____ 1. A collection of all the objects to be studied is a
   a. sample.
   b. population.
   c. variable.
   d. datum.

_____ 2. The highest level of measurement is
   a. nominal.
   b. ordinal.
   c. interval.
   d. ratio.

_____ 3. The level of measurement that can only be classified into groups is called
   a. nominal.
   b. ordinal.
   c. interval.
   d. ratio.

_____ 4. Rating a teacher as poor, fair, average, good, or superior would be what
         level of measurement?
         a. nominal
         b. ordinal
         c. interval
         d. ratio

_____ 5. A subset or part of the subjects to be studied is a
         a. sample.
         b. population.
         c. variable.
         d. datum.

_____ 6. I.Q.'s are an example of what level of measurement?
         a. nominal
         b. ordinal
         c. interval
         d. ratio

_____ 7. A sample that does not represent a population correctly is called:
         a. random.
         b. clustered.
         c. biased.
         d. unbiased.

_____ 8. Sampling that subdivides the population into subgroups is called:
         a. cluster.
         b. random.
         c. stratified.
         d. sequence.

_____ 9. Using preexisting groups in sampling is called:
         a. stratified.
         b. random.
         c. systematic.
         d. clusters.

_____ 10. If you sample every third item, you would be using:
         a. random sampling.
         b. sequence sampling.
         c. cluster sampling.
         d. systematic sampling.

_____ 11. Numbering all the items and then drawing numbers from a hat to determine
          which ones to test would be:
         a. random sampling.
         b. sequence sampling.
         c. cluster sampling.
         d. systematic sampling.

QUESTIONS:

CHAPTER 2
FREQUENCY DISTRIBUTIONS AND GRAPHS
Understanding Frequency Distributions and Graphs

Two main functions of descriptive statistics are summarizing and presenting data. The most common way to summarize data is in a frequency distribution. Charts and graphs are used to present data.

FREQUENCY DISTRIBUTIONS

A **frequency distribution** summarizes data by telling how many frequencies appear in each group or class. A **categorical frequency distribution** is used for nominal data and lists the categories and tells how many are in each category. Numerical data can presented in ungrouped or grouped frequency distributions. An **ungrouped frequency distribution** lists each number and the frequency for that number. A **grouped frequency distribution** gives several classes and the frequencies for each class. To decide whether to use an ungrouped or a grouped frequency distribution, find the range. The **range** is the highest number minus the lowest number in the data set. If the range is small, use an ungrouped frequency distribution. Below are examples of each type of frequency distribution.

Chart 1 Categorical: Grades for Elementary Statistics, Fall 1993

Grades	Frequencies (the number of students that made each grade)
A	3
B	4
C	7
D	2
F	1
	17

Chart 2 Ungrouped: Ages of Students in the 5th Grade at Stockton Elementary, 1993

Ages	Frequencies (the number of students for each age)
9	2
10	18
11	16
12	9
	45

Chart 3 Grouped: Annual Salaries for Franklin Factory Employees, 1993

Salaries ($000) Class limits	Frequencies (the number of employees in each salary class) f
15-19	12
20-24	13
25-29	27
30-34	18
35-39	10
	80

Grouped frequency distributions have parts besides the class limits and frequencies that can be found if the limits and the frequencies are given. The **class boundaries** are obtained by taking half of the distance from one upper class limit to the next upper class. Subtract this amount from each lower limit and add this amount to each upper limit. For Chart 3, half the distance from each lower limit to the next upper limit is .5. Subtracting .5 from each lower limit and adding .5 to each upper limit gives the following boundaries:

Class limits	f	Boundaries
15-19	12	14.5-19.5
20-24	13	19.5-24.5
25-29	27	24.5-29.5
30-34	18	29.5-34.5
35-39	10	34.5-39.5
	80	

Note that class limits do not overlap, but class boundaries always overlap. One upper class boundary is always the same as the next lower class boundary.

The **class width** is the difference between the upper and lower class boundary. The class width for the first class in Chart 3 would be $19.5 - 14.5 = 5$. The **midpoint** for each class is the the average of the lower and upper class limits. Midpoints are usually labeled "x" on the frequency distribution. The **cumulative frequencies** are found by writing down the first frequency and then adding the succeeding frequency to get the next cumulative frequency. Cumulative frequencies are labeled CF.

Chart 4 Complete grouped frequency distribution:
Annual Salaries for Franklin Factory Employees, 1993

Class limits	f	Boundaries	x	CF
15-19	12	14.5-19.5	17	12
20-24	13	19.5-24.5	22	25
25-29	27	24.5-29.5	27	52
30-34	18	29.5-34.5	32	70
35-39	10	34.5-39.5	37	80
	80			

RULES FOR FREQUENCY DISTRIBUTIONS

1. There should be between 5 and 20 classes.
2. The class width should be odd.
3. The class limits must not overlap.
4. The classes should be continuous.
5. Classes should include all the data.
6. The classes must be equal in width.

STEPS FOR MAKING A GROUPED FREQUENCY DISTRIBUTION

1. Find the highest and lowest value.
2. Find the the range. (highest − lowest)
3. Decide on the number of classes to use (between 5 and 15).

4. Class width $= \dfrac{\text{range}}{\text{number of classes}}$ (round up).

5. Take the lowest value (or slightly lower) for the first lower class limit. Add the class width to get the next lower class width. Keep adding the class width to get the rest of the lower class limits.
6. Find the upper class limits. If the data is rounded off to the units place, subtract one from the second lower class limit to get the first upper class limit. Then add the width to get the rest of the upper class limits. If the data is rounded off to the tenths place, subtract .1. If the data is rounded to the hundredths place, subtract .01 and so on.
7. Find the boundaries.
8. Tally the frequencies for each class.
9. Write down the frequencies.
10. Find the cumulative frequencies.

GRAPHS FOR A FREQUENCY DISTRIBUTION

There are three main types of graphs used to represent data that is in a frequency distribution. They are the histogram, the frequency polygon and the ogive. On all graphs and charts, equal spaces should always represent equal amounts. Graphs and charts should always be labeled and titled.

The **histogram** is a bar graph of a frequency distribution. The height of the bars represent the frequencies and the bottom of the bars are labeled with the class boundaries. Since the class boundaries overlap, the bars touch on the sides.

The **frequency polygon** is a line graph of a frequency distribution. The frequencies are plotted up the side and the midpoints are labeled along the bottom. The numbers written on the bottom should actually be the midpoints. The midpoints should be the same distance apart.

The **ogive** is a graph of the cumulative frequencies and the upper boundaries. The cumulative frequencies are plotted up the side and the upper class limits are labeled along the bottom. An example of the three graphs for the frequency distributions in Chart 4 are given below.

Histogram: Annual Salaries for Franklin Factory Employees, 1993

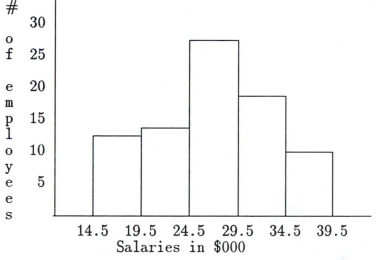

Frequency Polygon: Annual Salaries for Franklin Factory Employees, 1993

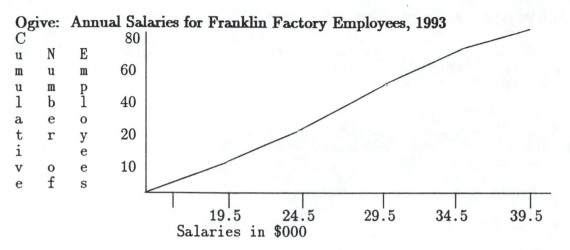

Ogive: Annual Salaries for Franklin Factory Employees, 1993

Cumulative Number Employees / Salaries in $000

Relative frequency graphs use the proportion for each group instead of the frequencies. The proportion for each group is found by dividing each frequency by the total number of frequencies. Relative frequency histograms, frequency polygons and ogives will look like the previous graphs, except the proportions will be graphed up the side instead of the frequencies.

OTHER GRAPHS AND CHARTS

Bar graphs use bars to represent how many are in each group shown. The bars can be horizontal or vertical, but the bars need to be the same width and the same space should be used between each bar. Time series graphs plot numbers over a period of time by putting the series of time along the bottom of the graph and the amount for each period of time as a point above that time and then connects the dots with a line to show the trend over time. Pie charts divide a circle into portions representing the percent of total for each category. Stem-and-leaf plots use the first digit (or digits) as the stem and the last digit as the leaf. Pictographs use a symbol or picture to represent a stated amount for each group. An example of each type of graph is given next.

Component Bar Graph: Enrollment at Williams Junior College (in 000)

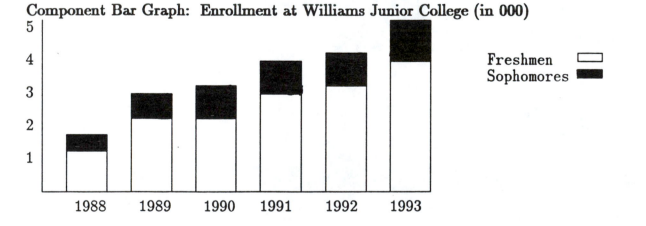

Freshmen
Sophomores

Time Series Graph: Total Enrollment at Williams Junior College (in 000)

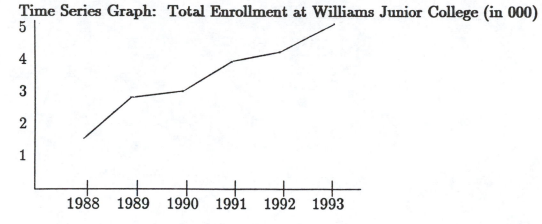

Pie Chart: Grades in Intermediate Algebra, Fall 1993, Williams Junior College

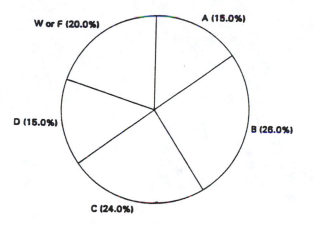

Stem-and-Leaf Plot: Ages of 25 Random Customers at a Record and Tape Store

Stem Leaf

Stem	Leaf
1	2, 3, 3, 4, 7, 8, 8, 9
2	0, 1, 2, 2, 3, 3, 4, 5, 8, 9
3	1, 2, 5, 6
4	3,8
5	5

Pictograph: Male Students in Homemaking at Franklin High School

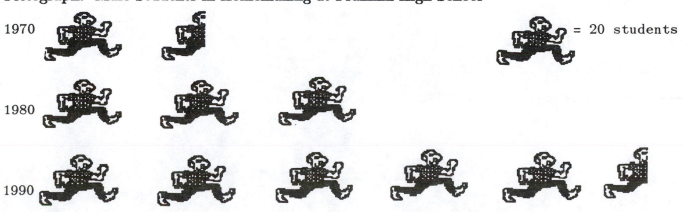

NOTES:

Checking Your Understanding

Complete this section before you go on. Write in the book. The answers are in the back of the book. Make a note of any questions that you wish to discuss with your instructor.

I. This frequency distribution has several errors. Describe at least four errors.

	f
0-10	6
10-30	3
40-50	9

II. Describe the errors in the bar graph.

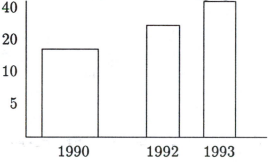

III. Select the correct answer and write the appropriate letter in the space provided.

____ 1. Histograms graph frequencies on the side and label _____ on the bottom of the graph.
   a. class limits
   b. class boundaries
   c. midpoints
   d. cumulative frequencies

____ 2. Frequency polygons plot frequencies on the side and label _____ on the bottom of the graph.
   a. class limits
   b. class boundaries
   c. midpoints
   d. cumulative frequencies

____ 3. Ogives plot cumulative frequencies on the side of the graph and label _____ on the bottom of the graph.
   a. upper class limits
   b. lower class limits
   c. upper boundaries
   d. lower boundaries

____ 4. A frequency distribution should have between ____ and ____ classes.
    a. 2 and 5
    b. 20 and 30
    c. 5 and 10
    d. 5 and 15

QUESTIONS:

Applying Your Understanding

STUDENT: In the preceding sections, you learned the concepts of making and reading frequency distributions and graphs, and checked your understanding of the concepts. In this section, you will apply your understanding of the concepts. Study each example carefully and then try to work the following exercise. If you have any problems, see your instructor. The answers to the exercises are in the back of the book.

Example 1--Making a frequency distribution.

Make a frequency distribution from the numbers below.

Rent for One Bedroom Apartments in 20 Different Complexes

200	185	325	320	310
290	250	285	190	290
180	225	210	225	235
235	245	205	230	260

Solution

Step 1: high = 325   low = 180

Step 2: Range = 325 − 180 = 145

Step 3: Any number between 5 and 20 is acceptable for the number of classes, but for a data set this small a fairly small number (between 5 and 8) would be best. The number 6 is arbitrarily picked for this problem.

Step 4: Class width = $\frac{145}{6} = 24.17$ Round up to 25. (Round up goes to the next higher whole number.)

Step 5: 180 or a little lower is an acceptable starting point for the first lower class limit. Since the width is 25, 175 might be used as a starting point. Add 25 to find all the lower class limits.

Lower class limits

175   (add 25 to get the next lower limit)
200   (keep adding 25 to get the next limit)
225
250
275
300
325
350   (not needed since the highest number is 325)

This would give 7 classes which is acceptable since 6 was picked arbitrarily, but if you want to have exactly 6 classes, you could start at 180.

Lower class limits

180
205
230
255
280
305
330   (not needed)

Either set of lower limits would be acceptable (there is not just one correct frequency distribution). This time the second set is chosen to keep 6 classes.

Step 6: Subtract 1 from 205 (second class limit) to get the first upper class limit of 204. Then add the width, 25 to get the rest of the upper class limits.

Class limits
180 - 204
205 - 229
230 - 254
255 - 279
280 - 304
305 - 329   NOTE: The class limits do not overlap.

Step 7: Subtract .5 from each lower class limit. Add .5 to each upper class limit to get the boundaries.

Class limits	Boundaries
180 - 204	179.5 - 204.5
205 - 229	204.5 - 229.5
230 - 254	229.5 - 254.5
255 - 279	254.5 - 279.5
280 - 304	279.5 - 304.5
305 - 329	304.5 - 329.5

NOTE: The boundaries do overlap.

Step 8: Put a mark beside the appropriate class for each number in the data set.

Class limits	Boundaries	Tally
180 - 204	179.5 - 204.5	////
205 - 229	204.5 - 229.5	////
230 - 254	229.5 - 254.5	/////
255 - 279	254.5 - 279.5	/
280 - 304	279.5 - 304.5	///
305 - 329	304.5 - 329.5	///

Step 9: Write down the frequencies.

Class limits	Boundaries	Tally	f
180 - 204	179.5 - 204.5	////	4
205 - 229	204.5 - 229.5	////	4
230 - 254	229.5 - 254.5	/////	5
255 - 279	254.5 - 279.5	/	1
280 - 304	279.5 - 304.5	///	3
305 - 329	304.5 - 329.5	///	3
			20

Step 10: Find the cumulative frequencies by adding each frequency to the previous cumulative frequency.

Class limits	Boundaries	Tally	f	cf	
180 - 204	179.5 - 204.5	////	4	4	(1st f)
205 - 229	204.5 - 229.5	////	4	8	(4 + 4)
230 - 254	229.5 - 254.5	/////	5	13	(8 + 5)
255 - 279	254.5 - 279.5	/	1	14	
280 - 304	279.5 - 304.5	///	3	17	
305 - 329	304.5 - 329.5	///	3	20	
			20		

(The numbers in parentheses are not part of the distribution.)

Exercise 1

Make a frequency distribution for the ages of 30 students given below.

18	19	29	17	27
21	21	36	22	24
24	23	35	23	33
26	25	18	26	28
28	27	22	29	19
18	31	25	30	33

Example 2--Making a frequency distribution with data containing decimals.

Make a frequency distribution from the numbers given below.
Braking time in seconds for a sample of 23 year old males

.41	.86	.67	.63	.69
.62	.93	.58	.56	.73
.85	.58	.76	.83	.87
.90	.94	.87	.86	.98
.51	.89	.48	.65	.49

Solution

Step 1: high = .98   low = .41

Step 2: Range = .98 − .41 = .57

Step 3: Arbitrarily pick 7 classes.

Step 4: Class width = $\frac{.57}{7}$ = .0814   Round up to the same decimal place as the data, so use .09 as class width.

Step 5: Since the data is in hundredths, the limits need to be in hundredths. Use .41 as the first class limit. Add .09 to get the lower class limits.
Lower class limits
.41
.50
.59
.68
.77
.86
.95
1.04   (not needed)

Step 6: Since the data is in hundredths, subtract .01 from the second lower class limit to get the first upper class limit. Then add the class width .09 to get each of the next upper class limits.
Class limits
.41 - .49
.50 - .58
.59 - .67
.68 - .76
.77 - .85
.86 - .94
.95 -1.03

Step 7: Since the data is in hundredths, subtract .005 (half of .01) from each lower class limit and add .005 to each upper class limit to get the boundaries.

Class limits	Boundaries
.41 - .49	.405 - .495
.50 - .58	.495 - .585
.59 - .67	.585 - .675
.68 - .76	.675 - .765
.77 - .85	.765 - .855
.86 - .94	.855 - .945
.95 -1.03	.945 -1.035

Step 8: Tally.

Class limits	Boundaries	Tally
.41 - .49	.405 - .495	///
.50 - .58	.495 - .585	////
.59 - .67	.585 - .675	////
.68 - .76	.675 - .765	///
.77 - .85	.765 - .855	//
.86 - .94	.855 - .945	////////
.95 -1.03	.945 -1.035	/

Step 9: Write down the frequencies.

Class limits	Boundaries	Tally	f
.41 - .49	.405 - .495	///	3
.50 - .58	.495 - .585	////	4
.59 - .67	.585 - .675	////	4
.68 - .76	.675 - .765	///	3
.77 - .85	.765 - .855	//	2
.86 - .94	.855 - .945	////////	8
.95 -1.03	.945 -1.035	/	1
			25

Step 10: Find the cumulative frequencies.

Class limits	Boundaries	Tally	f	cf
.41 - .49	.405 - .495	///	3	3
.50 - .58	.495 - .585	////	4	7
.59 - .67	.585 - .675	////	4	11
.68 - .76	.675 - .765	///	3	14
.77 - .85	.765 - .855	//	2	16
.86 - .94	.855 - .945	////////	8	24
.95 -1.03	.945 -1.035	/	1	25
			25	

Exercise 2

Make a frequency distribution from the GPA's listed below.

2.36	1.56	2.48	2.69
3.89	2.98	3.16	1.57
2.14	2.76	3.35	1.82
1.78	2.64	3.56	2.36

Example 3--Making a histogram.

Make a histogram from the frequency distribution of 80 college professors and their years of teaching experience.

Class limits	f	Boundaries
0 – 4	16	– .5 – 4.5
5 – 9	21	4.5 – 9.5
10 –14	12	9.5 – 14.5
15 –19	11	14.5 – 19.5
20 –24	10	19.5 – 24.5
25 –29	8	24.5 – 29.5
30 –34	2	29.5 – 34.5
	80	

Solution

Label each boundary along the bottom of the graph.

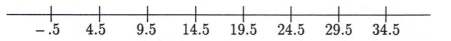

Mark the side with any interval as long as equal spaces represent equal amounts.

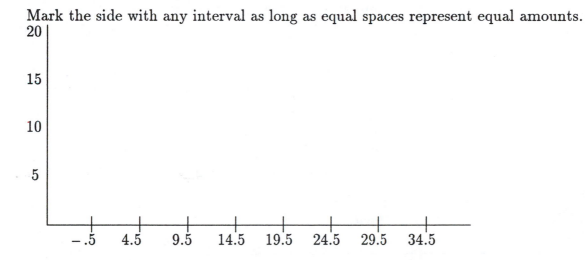

Draw the bars making the bars touch on the sides. Label and title the graph.

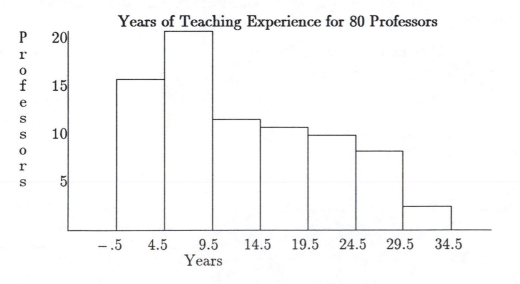

Years of Teaching Experience for 80 Professors

Exercise 3

Make a histogram for the following frequency distribution of grades in a college algebra class.

Class limits	f
40 - 49	2
50 - 59	3
60 - 69	4
70 - 79	8
80 - 89	5
90 - 99	2
	24

Example 4--Making a frequency polygon.

Make a frequency polygon for the frequency distribution in Example 3.

Solution

The midpoints are needed to make a frequency polygon. Average the lower and upper class limits for each class to get the midpoints. Label the midpoint column "x."

Class limits	f	x	
0 - 4	16	2	$(0 + 4) \div 2$
5 - 9	21	7	$(5 + 9) \div 2$ or add the class width 5 to each
10-14	12	12	midpoint to get the next midpoint
15-19	11	17	
20-24	10	22	
25-29	8	27	
30-34	2	32	
	80		

Label the midpoints along the bottom of the graph keeping an equal space between each and leaving an empty space at each end.

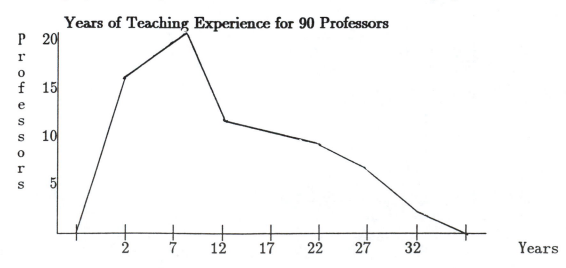

Put a point for each frequency above its midpoint. Connect the points and connect the end of the graph to the spaces at each end. Label and title the graph.

Years of Teaching Experience for 90 Professors

Exercise 4

Make a frequency polygon for the frequency distribution in Exercise 3.

Example 5--Making an ogive.

Make an ogive for the frequency distribution in Example 3.

Solution

Find the cumulative frequencies by writing the first frequency and adding each successive frequency to get the rest of the cumulative frequencies.

Class limits	f	Boundaries	cf
0 – 4	16	–.5 – 4.5	16
5 – 9	21	4.5 – 9.5	37
10 –14	12	9.5 – 14.5	49
15 –19	11	14.5 – 19.5	60
20 –24	10	19.5 – 24.5	70
25 –29	8	24.5 – 29.5	78
30 –34	2	29.5 – 34.5	80
	80		

Label the upper boundaries along the bottom of the graph leaving blank space at the first. Mark the sides with any interval as long as each space is the same. Go at least as far as 80 (the highest cumulative frequency).

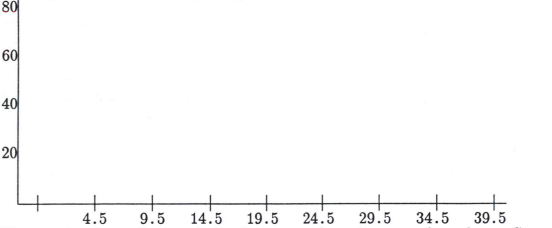

Plot a point for each cumulative frequency above its upper boundary. Connect the points with a line. On the left side, draw a line to the bottom of the graph to the blank space. Label and title the graph.

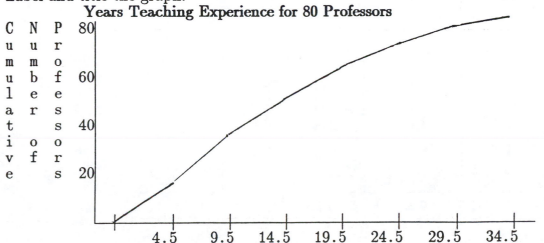

Exercise 5

Make a ogive for the frequency distribution in Exercise 3.

Example 6--Making relative frequency graphs.

Make a relative frequency histogram, frequency polygon, and ogive for the frequency distribution in Example 3.

Solution

Find the proportions for each group by dividing each frequency by the total number of frequencies.

Class limits	f	Boundaries	Proportion	
0 - 4	16	-.5 - 4.5	.20	(16 ÷ 80 = .2)
5 - 9	21	4.5 - 9.5	.26	(21 ÷ 80 = .2625)
10 -14	12	9.5 - 14.5	.15	(12 ÷ 80 = .15)
15 -19	11	14.5 - 19.5	.14	(11 ÷ 80 = .1375)
20 -24	10	19.5 - 24.5	.13	(10 ÷ 80 = .125)
25 -29	8	24.5 - 29.5	.10	(8 ÷ 80 = .1)
30 -34	2	29.5 - 34.5	.03	(2 ÷ 80 = .025)
	80		1.01	

Keep all the numbers rounded off to the same decimal place. The total must be exactly 1, so adjust a value to get exactly 1.

Class limits	f	Boundaries	Proportion	
0 - 4	16	-.5 - 4.5	.20	(16 ÷ 80 = .2)
5 - 9	21	4.5 - 9.5	.26	(21 ÷ 80 = .2625)
10 -14	12	9.5 - 14.5	.15	(12 ÷ 80 = .15)
15 -19	11	14.5 - 19.5	.14	(11 ÷ 80 = .1375)
20 -24	10	19.5 - 24.5	.13	(10 ÷ 80 = .125)
25 -29	8	24.5 - 29.5	.10	(8 ÷ 80 = .1)
30 -34	2	29.5 - 34.5	.02	(2 ÷ 80 = .025)
	80		1.00	

A relative frequency histogram labels the boundaries along the bottom and the proportions along the side.

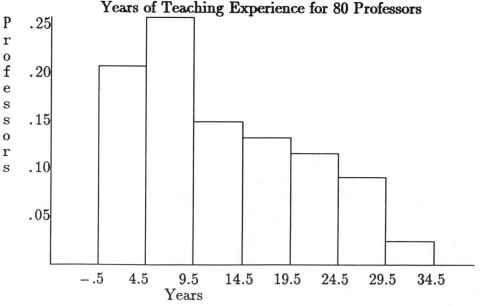

Years of Teaching Experience for 80 Professors

A relative frequency polygon labels the midpoints along the bottom and the proportions up the side.

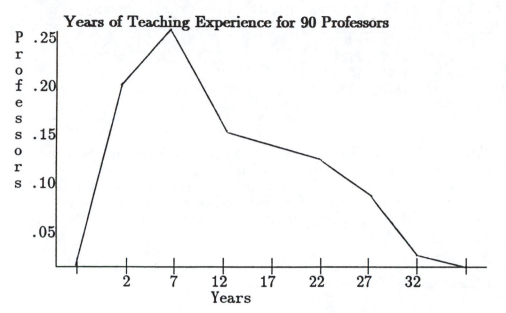

A relative frequency ogive labels the upper boundaries along the bottom and the cumulative proportions up the sides. To find the cumulative proportions, write down the first proportion and then add each successive proportion.

Class limits	f	Boundaries	Proportion	Cumulative proportion
0 – 4	16	− .5 – 4.5	.20	.20
5 – 9	21	4.5 – 9.5	.26	.46
10 –14	12	9.5 – 14.5	.15	.61
15 –19	11	14.5 – 19.5	.14	.75
20 –24	10	19.5 – 24.5	.13	.88
25 –29	8	24.5 – 29.5	.10	.98
30 –34	2	29.5 – 34.5	.02	1.00
	80		1.00	

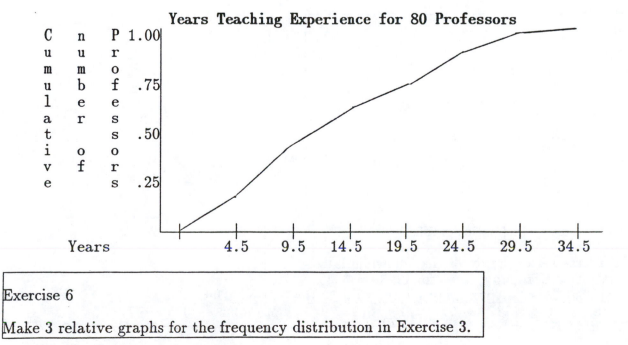

<hr>

Exercise 6

Make 3 relative graphs for the frequency distribution in Exercise 3.

Example 7--Making a component bar graph.

Make a component bar graph for the following data on the average amount of money spent by a family of four.

Gasoline		Groceries	
1991	$70	1991	$495
1992	$79	1992	$520
1993	$110	1993	$560

Solution

Label three bars 1991, 1992, 1993. Keep the bars the same width and the space between the bars equal. Label up the sides with any equal intervals making sure to go at least as high as $670 (the total for 1993). Draw a bar for each year up the the height for the gasoline spent. Draw another box on top of each gasoline box adding the grocery amount to see how far up to make the bar. Label and title the graph. Color in either the gasoline or grocery box.

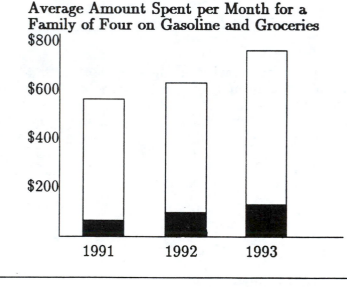

Average Amount Spent per Month for a Family of Four on Gasoline and Groceries

Groceries

Gasoline

Exercise 7

Draw a component bar graph for the following data.

Sales for a Women's Shoe Store (in $000)

	1991	1992	1993
Shoes	42	49	54
Purses	12	15	10
Hosiery	8	12	14

Example 8--Making a time series graph.

Make a time series graph for the following data.
Robert's Toy Company, Sales for 1993 (in $000)

Jan.	Feb.	Mar.	Apr.	May	Jun.	Jul.	Aug.	Sep.	Oct.	Nov.	Dec.
17	18	20	28	15	21	19	14	12	22	32	57

Solution

Label the months along the bottom. Mark equal intervals up the side at least as far as 57, the highest value. Put a point above each month across from the amount sold for that month. Connect the points label and title the graph.

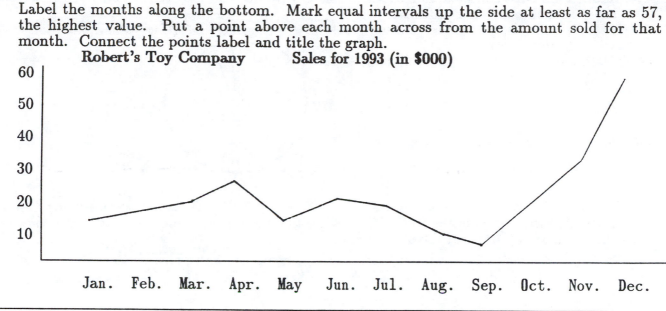

Robert's Toy Company   **Sales for 1993 (in $000)**

Exercise 8

Make a time series graph for the following data for Doctorates awarded in Physical Science.

1987	1988	1989	1990	1991	1992	1993
2500	2600	2750	2980	3000	3300	3490

Example 9--Making a pie graph.

Make a pie graph for the following data.

Franklin Factory, Expenses 1993

Salaries	58962
Maintenance	212317
Production Costs	790862
Supplies	170916
Taxes	27871

Solution

Pie charts use percents of the total so first add all the values and divide each number by the total amount to get the percent of the total for each amount. Make the sure the total percent is exactly 100%. Find the cumulative percents by adding each percent to the previous percent.

		Percents	Cumulative Percents
Salaries	58962	4.7%	4.7%
Maintenance	212317	16.8%	21.5%
Production Costs	790862	62.7%	84.2%
Supplies	170916	13.6%	97.8%
Taxes	27871	2.2%	100%
	1260928		

The text explains how to make a pie chart using a compass and a protractor, but if you do not have these available, you can still make a pie chart. Use any round object to draw a circle on your paper. Put four marks at equal intervals on the circle. Since the circle represents 100%, each mark represents 25%. Label the four marks 0%, 25%, 50% and 75%. Put four marks at equal spacing in between each of the first four marks. Each of these marks now represents 5%. Draw a line from 0% to the center of the circle. Now draw another line from the center to 4.7% (or as close as you can tell). This piece of the pie represents the percentage for salaries.

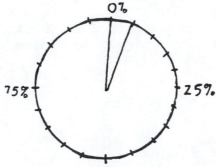

Draw a line from the center to 21.5% (4.7% + 16.8%) to the show the piece for buildings. Draw line from the center to 84.2% to show the piece for production costs. Put the rest of the pieces on. Title and label the graph.

Taxes (2.2%)   Salaries (4.7%)
Supplies (13.6%)
Maintenance (16.8%)
Production (62.7%)

Exercise 9
Make a pie chart for the data given below.
Franklin Factory, Sales in %000, 1993

Gadgets	Gidgets	Widgets	Thing-a-ma-jigs	What-cha-ma-call-its
278	324	478	515	314

Example 10–Making a stem-and-leaf plot.
   Make a stem-and-leaf plot for the following numbers.

215	239	212	245	226	228	246	213	247	225
236	223	221	248	237	242	218	236	232	238

Solution
   Step 1: Arrange the data in order: 212 213 215 218 221 223 225 226 228
         232 236 237 238 239 242 245 246 247 248
   Step 2: Separate the data according to classes. Since the first digit for all of these numbers is 2, use the second digit to separate into classes.
   212 213 215 218   221 223 225 226 228
   232 236 236 237 238 239   242 245 246 247 248
   Step 3: Use the first 2 digits for the leading digit (or stem) and list all of the last digits in order for the trailing digit (or leaf).

Leading digit	Trailing digit
21	2 3 5 8
22	1 3 5 6 8
23	2 6 6 7 8 9
24	2 5 6 7 8

Exercise 10
Make a stem-and-leaf plot for the following numbers:

48	63	78	42	49	51	65	56	53	41	47	73	79	58	62

Practice Test

1. Make a frequency distribution of the ages that 25 randomly selected smokers started smoking:

26	26	25	17	16	16	14	17	21	16	16	18	17
15	15	19	16	17	22	15	19	17	16	21	16	

   Use the following frequency distribution for problems 2, 3, 4, and 5.

   Average number of miles driven a day for 30 commuters:

Miles	Number of commuters
Stated limits	f
0-24	4
25-49	10
50-74	11
75-99	5

2. Make a histogram.

3. Make a frequency polygon.

4. Make an ogive.

5. Make a relative histogram, frequency polygon and ogive.

6. Make a bar graph of types of albums sold at Cedar Point Record Store in May, 1993

Rock	Country and western	Easy listening	Pop	Classical	Jazz
215	172	62	89	64	118

7. Make a time series graph for the sales of Cedar Point Record Store.

Year	Sales (in $000)
1989	28
1990	32
1991	30
1992	35
1993	38

8. Make a stem-and-leaf plot.

10	20	25	27	28	12	13	25	38
29	36	32	31	42	43	41	37	17

CHAPTER THREE
DATA DESCRIPTION
Understanding Data Description

Introduction

Data can be described by three different measures. **Measures of central tendency** are the averages and tell about the middle of the numbers. **Measures of variation** tell if the numbers are close together or spread far apart. **Measures of position** tell the relative position of a number in comparison with the rest of the numbers.

Data from a population are called **parameters**. Data from a sample are called **statistics**. In some of the following measures, it will be important to note if the data comes from a population or from a sample.

Measures of Central Tendency

The **mean** is the sum of all the numbers in a data set divided by the total number of values. If the data is from a sample, $\bar{x}$ is used of represent the mean and n is used to represent the total number of values in the data set. If the data is from a population, μ is used to represent the mean and N is used to represent the total number of values in the data set. (μ is the Greek letter mu.) The method for finding the mean is the same for a population or a sample, but it is important to label the mean correctly because the formulas for other measures are different depending on whether the data is from a population or a sample. The capital Greek letter sigma, Σ, is called a summation symbol in formulas and means to add up all the values.

Mean	
Sample	Population
$\bar{x} = \frac{\Sigma x}{n}$	$\mu = \frac{\Sigma x}{N}$

If the grades for a sample of six students are 96, 92, 81, 75, 61, and 48, then the mean is:

$$\bar{x} = \frac{96 + 92 + 81 + 75 + 61 + 48}{6} = \frac{453}{6} = 75.5$$

The mean for ungrouped data in a frequency distribution is found by multiplying the values by the frequency for each set of numbers, adding all the products, and dividing by the total number of frequencies.

Mean for ungrouped frequency distribution	
Sample	Population
$\bar{x} = \frac{\Sigma fx}{n}$	$\mu = \frac{\Sigma fx}{N}$

The wages paid at a fast food restaurant and the number of people making each wage are given next.

Wage (x)	Number of employees (f)	fx
$3.90	20	78.00
$4.12	12	49.44
$4.75	5	23.75
	37	151.19

$$\mu = \frac{151.19}{37} = 4.086 \text{ (or \$4.09 since 4.086 does not make any sense in dollars).}$$

To find the mean for data in a frequency distribution, multiply the frequency of each group by the midpoint of the class, total the products and divide by the total number frequencies.

Mean for grouped frequency distribution

Sample   Population   where x is the midpoint of each group

$\bar{x} = \dfrac{\sum fx}{n}$ $\mu = \dfrac{\sum fx}{N}$

The rent for a sample of one bedroom apartments is given below.

Rent

Class Limits	f	x	fx
180 - 204	4	192	768
205 - 229	5	217	1085
230 - 254	4	242	968
255 - 279	1	267	267
280 - 304	3	292	876
305 - 329	3	317	951
	20		4915

$\bar{x} = \dfrac{4915}{20} = 245.75$ The mean rent is \$245.75

Median

The **median** is the middle number. Half of the values in the data set are smaller than the median and half of the values are larger than the median. To find the median, first put the numbers in order, then find the middle number. If there is an odd number of values, the number in the middle will be the median. If there is an even number of values, then the average of the two numbers in the middle will be the median. The symbol for median is **MD**.

The ages of 11 employees in a convenience store are: 24   26   32   45   18   21   53   19   28   24   38.   To find the median , put the numbers in order and find the middle number. 18   19   21   24   24   26   28   32   38   45   53   MD = 26   Half of the ages are below 26, half are above 26.

The grades for 12 students on a statistic test are: 81   75   98   72   86   61   58   90   85   74   78   74. To find the median, put the numbers in order. Since there are an even number of values, the median is the average of middle two numbers. 58   61   72   74   74   75   78   81   85   86   90   98   MD $= \dfrac{75 + 78}{2} = \dfrac{153}{2} = 76.5$ Half of the grades were above 76.5 and half were below.

To find the median for an ungrouped frequency distribution, find $\frac{n}{2}$ to find the position of the middle number. Then look down the cumulative frequency column   to find the first number that is $\frac{n}{2}$ or larger. The class value for that cumulative frequency is the median. The number of students absent in a mathematics class for one month are given below.

Number of students absent	Days (f)	cf
0	2	2
1	1	3
2	2	5
3	10	15 * median class md = 3
4	3	18
5	3	21
6	1	22
	22	

$\frac{n}{2} = \frac{22}{2} = 11$

The median number of absences was 3. Half of the students had less than 3 absences, half had more than 3 absences.

To find the median for a grouped frequency distribution, use the formula:

$$MD = \frac{\frac{n}{2} - cf}{f} (w) + L_m$$

n = total number of frequencies
cf = cumulative frequency of the class above the median class
f = frequency of the median class
w = width of the median class
L_m = lower boundary of the median class

To find the median class, find $\frac{n}{2}$ and look down the cumulative frequency column until you find the first number that is $\frac{n}{2}$ or larger. That will be the median class. The median for the annual salaries for employees for Franklin Factory, 1993, is found below (in $000).

Class limits	f	Boundaries	cf
15 - 19	12	14.5 - 19.5	12
20 - 24	13	19.5 - 24.5	25
25 - 29	27	24.5 - 29.5	52 * median class
30 - 34	20	29.5 - 34.5	72
35 - 39	18	34.5 - 39.5	90
40 - 44	10	39.5 - 44.5	100
	100		

$MD = \frac{50 - 25}{27} (5) + 24.5 = 29.130$ Half of the salaries were above $29,130.

Mode

The **mode** is the number that appears the most often in the data set. There might be one or two or more modes or there may not be a mode for every data set. The ages of 11 employees in a convenience store are: 24 26 32 45 18 21 53 19 28 24 38. The mode for the ages is 24 since 24 is the number that appears the most.

The mode for an ungrouped frequency distribution is the value that has the most frequencies. The number of absences for a class were:

Number of students absent	Days
0	2
1	1
2	2
3	10 * most frequencies
4	3
5	3
6	1

The mode is 3 absences since there are 10 days that 3 students were absent.

The mode for a grouped frequency distribution is the midpoint of the class with the most frequencies. The mode for salaries for employees for Franklin Factory, 1993 is found below (in $000).

Class limits	f	Midpoints(x)
15 - 19	12	17
20 - 24	13	22
25 - 29	27*	27
30 - 34	20	32
35 - 39	18	37
40 - 44	10	42
	100	

The class with 27 frequencies is the modal class so the mode is 27, the midpoint of the class.

$$\text{midrange} = \frac{\text{highest value} + \text{lowest value}}{2}$$

The ages for the employees at a convenience store were: 24   26   32   45   18   21   53   19   28   24   38.

The midrange is $\frac{18 + 53}{2} = \frac{71}{2} = 35.5$.

Comparison of Mean, Median, and Mode

The mean uses all the values, so it can be affected by a few very large or very small numbers. If the mean is quite a bit different from the median or the mode, sometimes it is not a very good representation of the data. The mean is unique, but cannot be found for categorical data or for open-ended frequency distributions. The median does not use all the values so it is less affected than the mean by a few large or small numbers. The median is unique. and can be found for open-ended frequency distributions. The mode can be found for nominal data but is not unique. There may be more than one mode or there may not be a mode. If the mean, median, and the mode are all equal, the frequency distribution is **symmetrical**. If the mean is larger than the median, the distribution is **positively skewed**. If the mean is smaller, then the distribution is **negatively skewed**.

Measures of Variation

Measures of variation are used to tell if the numbers in the data are close together or spread far apart. The **range** is the highest number minus the lowest number in the data set and is a measure of variation. The **variance** and **standard deviation** are also used to show the spread of the numbers. The standard deviation is more commonly used and shows the average amount the numbers are deviating from the mean. **The standard deviation has to be compared to the mean.** If the standard deviation is large compared to the mean, then the numbers are not close together. If the standard deviation is relatively small, then the numbers are close together and there is not much deviation. There are different formulas for population and sample variances and standard deviations. For a sample, the variance is represented by s^2 and the standard deviation is represented by s. For a population, the variance is σ^2 and the standard deviation is σ (the small case Greek letter sigma).

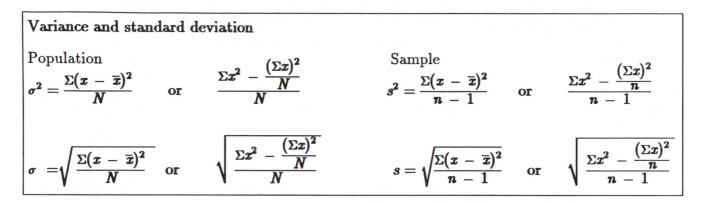

Variance and standard deviation

Population

$$\sigma^2 = \frac{\Sigma(x - \bar{x})^2}{N} \quad \text{or} \quad \frac{\Sigma x^2 - \frac{(\Sigma x)^2}{N}}{N}$$

Sample

$$s^2 = \frac{\Sigma(x - \bar{x})^2}{n - 1} \quad \text{or} \quad \frac{\Sigma x^2 - \frac{(\Sigma x)^2}{n}}{n - 1}$$

$$\sigma = \sqrt{\frac{\Sigma(x - \bar{x})^2}{N}} \quad \text{or} \quad \sqrt{\frac{\Sigma x^2 - \frac{(\Sigma x)^2}{N}}{N}}$$

$$s = \sqrt{\frac{\Sigma(x - \bar{x})^2}{n - 1}} \quad \text{or} \quad \sqrt{\frac{\Sigma x^2 - \frac{(\Sigma x)^2}{n}}{n - 1}}$$

The formulas for variance and standard deviation for a sample in a frequency distribution are given next. The x in the formula represents the midpoints for a grouped frequency distribution or the class value for an ungrouped frequency distribution.

Variance and standard deviation for frequency distribution

Sample

$$s^2 = \frac{\Sigma fx^2 - \frac{(\Sigma fx)^2}{n}}{n - 1} \qquad s = \sqrt{\frac{\Sigma fx^2 - \frac{(\Sigma fx)^2}{n}}{n - 1}}$$

The **coefficient of variation** makes it easier to tell if a standard deviation is large or small by comparing the standard deviation to the mean and it allows comparison of standard deviations that come from data sets with different means. The coefficient of variation is represented by CV.

Coefficient of Variation

Sample   Population

$$\text{CV} = \frac{s}{\bar{x}}(100\%) \qquad\qquad \text{CV} = \frac{\mu}{\sigma}(100\%)$$

The **standard** or **z-score** allows comparisons of values that come from groups with different means and standard deviations.

z-score

Sample   Population

$$z = \frac{(x - \bar{x})}{s} \qquad\qquad z = \frac{(x - \mu)}{\sigma}$$

Distribution Spreads

Chebyshev's Theorem uses the standard deviation to determine what percent of a data set will fall in a certain range.

Chebyshev's Theorem: The proportion of values from a data set that will fall within k standard deviations of the mean will be at least $1 - \frac{1}{k^2}$, where k is a number greater than one.

Chebyshev's theorem can be applied to any distribution, but when a distribution is bell-shaped or normal, the **Empirical Rule** can be used.

Empirical Rule:
Approximately 68% of the data values will fall within one standard deviation of the mean.
Approximately 95% of the data values will fall within two standard deviations of the mean.
Approximately 99.7% of the data values will fall within three standard deviations of the mean.

Measures of Position

Measures of position compare the location of a value in a data set in relation to the other values. The most common measures of position are percentiles, deciles, and quartiles. **Percentiles** can be found for any percent from 1 to 99 and are denoted by P with a subscript for the desired percentile. For example, P_{10} is the tenth percentile and is the number that is larger than 10% of the other values. **Deciles** divide the data set into tenths and can be found for 1 through 9. Deciles are written as D with a subscript. D_3 is the third decile and is the value that is larger than three tenths of the other values. **Quartiles** divide the values into fourths and can be found for 1 and 3. Q_1 is the first quartile and is the value that is larger than one fourth of the other numbers. The second quartile would be two fourths or one half, so it is the same as the median and is not called Q_2.

Cumulative Percentage Graph, Franklin Factory Salaries, 1993

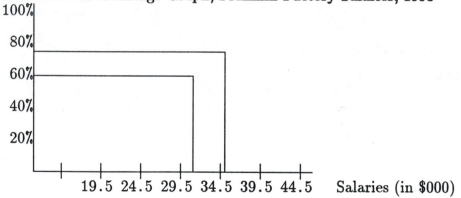

To find the percentile for a given value or to find the value for a particular percentile for a frequency distribution, a cumulative percentage graph is used. A cumulative percentage is drawn like a ogive by labeling the upper boundaries along the bottom of the graph, but the cumulative percentages for each group are plotted up the side.

Cumulative Percentage Graph, Franklin Factory Salaries, 1993

To find a percentile for a given value, look up from the approximate place of the value on the bottom of the graph until you hit the graph and then look across from that place on the graph until you can read the percentile on the left side of the graph. The percentile for $36,000 would be approximately 77%, so 77% of the salaries were below $36,000. To find the value for a certain percentile, look across from the desired percentile until you hit the graph and then look down until you can read the value off the bottom of the graph. The value for 60% would be approximately $32,000.

To find the percentile for data that is not in a frequency distribution, a formula is used.

$$\text{Percentile} = \frac{\text{number of values below the given value} + .5}{\text{total number of values in the data set}}(100\%)$$

Steps for Finding a Value for a Given Percentile
1. Arrange the data in order from smallest to largest.
2. Compute C.

$$C = \frac{n(\text{Percentile})}{100} \quad \text{where } n \text{ is the total number of values}$$

3. If C is not a whole number, round up to the next whole number, then the value is the number in that position looking at the values in order. If C is a whole number, then the value is the average of the numbers in the C and the $C + 1$ position.

Box and Whisker Plots

A **box and whisker plot** graphs five values of the data set on a number line. The five values are:
1. The lowest value in the data set
2. The lower hinge
3. The median
4. The upper hinge
5. The highest value of the data set

A box is drawn from the lower hinge to the upper hinge and lines are drawn from the box to the highest and lowest value. The **lower hinge** is the median of all values less than or equal to the median when the data set has an odd number of values, or the median of all values less than the median when the data set has an even number of values. The **upper hinge** is the median of all values greater than or equal to the median when the data set has an odd number of values, or the median of all values greater than the median when the data set has an even number of values.

To make a box and whisker plot for the age of 11 employees at a convenience store, first you need to find the highest value, the lowest value, the midpoint, upper and lower hinges.

24   26   32   45   18   21   53   19   28   24   38

highest value = 53   lowest value = 18   MD = 26   upper hinge = 35   lower hinge = 22.5

The box and whisker plot shows that the data is not symmetrical and that the data is positively skewed since the whisker is longer on the right.

NOTES:

Checking Your Understanding

Complete this section before you do the exercises to make sure you understand the concepts. Write in the book. The answers are in the back of the book. Make a note of any questions that you wish to discuss with your instructor.

I. A group of numbers has a mean of 375, median of 280, and a mode of 250. Tell which number you would use to represent the data and why.

II. A shopper compared prices at several stores for the same item and found a median price of $2.68. Explain what a median of $2.68 means.

III. A class had an average of 69.8 on the first test, with a standard deviation of 6.2. Explain 6.2.

IV. One class has an average of 69.8 on the first test with a standard deviation of 6.2. Another class has an average of 70.4 with a standard deviation of 13.4. Compare the performances of the two classes.

V. A student scored 72% on a test, and was in the 85th percentile. Explain these two numbers.

VI. Select the correct answer and write the appropriate letter in the space provided:

____1. The only measure of central tendency that can be found for nominal data is the
    a. mean.
    b. median.
    c. mode.
    d. midrange.

____2. Half of the numbers in any data set will be larger than the
    a. mean.
    b. median.
    c. mode.
    d. midrange.

____3. The number that appears most frequently in a data set is the
    a. mean.
    b. median.
    c. mode.
    d. midrange.

_____4. The measure of central tendency that is most affected by a few large or small numbers is
     a. mean.
     b. median.
     c. mode.
     d. midrange.

_____5. One fourth of the scores in any data set are below the
     a. first quartile.
     b. second quartile.
     c. third quartile.
     d. fourth quartile.

QUESTIONS:

Applying Your Understanding

STUDENT: In the preceding section, you learned the concepts of the measures of central tendencies, measures of variation, and the measures of position, and checked your understanding of the concepts. In this section, you will apply your understanding of the concepts. Study each example carefully and then try to work the following exercise. If you have any problems, see your instructor. The answers to the exercises are in the back of the book.

Example 1--Finding the mean.

Find the mean cost of this sample of speeding tickets. $25 $30 $65 $70 $40 $55 $60 $35 $50 $70 $95

Solution

Add all the values and divide by the total number of values, 11. Label the mean $\bar{x}$ since this is a sample.

$$\bar{x} = \frac{25 + 30 + 65 + 70 + 40 + 55 + 60 + 35 + 50 + 70 + 95}{11} = \frac{595}{11} = 54.09$$

The mean cost of these speeding tickets is $54.09.

> **Exercise 1**
>
> Find the mean cost of a three-item pizza for these sample prices. $7.95 $9.98 $5.58 $4.99 $10.75 $6.25 $8.88 $7.63 $8.50

Example 2--Finding the mean for an ungrouped frequency distribution.

Tickets for a concert were priced at different amounts depending on the seating and how far in advance the tickets were purchased. For the amounts paid and how many at each price given below, find mean cost of the tickets.

Price (x)	Number of tickets (f)
10	400
15	300
20	450
25	200
50	50

Solution

Multiply each price by the number of tickets sold at that price and total the products. Total the number of tickets sold. Divide the total of the products by the total number of tickets sold.

x	f	fx
10	400	4000
15	300	4500
20	450	9000
25	200	5000
50	50	2500
	1400	25,000

Label the mean price μ since these were all the tickets sold. $\mu = \dfrac{25,000}{1400} = \$17.86$

Exercise 2
Find the mean age for children in a fourth grade class.

Ages	Number of students
8	2
9	10
10	8
11	5

Example 3--Finding the mean for a grouped frequency distribution.

Find the mean years teaching experience for the sample of 80 college professors.

Class limits	f
0 - 4	16
5 - 9	21
10 - 14	12
15 - 19	11
20 - 24	10
25 - 29	8
30 - 34	2

Solution

Find the midpoints by adding the lower and upper limit for each class and dividing by 2. Multiply each frequency by its midpoint and total the products. Total the frequencies. Divide the total of the products by the total of the frequencies.

Class limits	f	x	fx
0 - 4	16	2	32
5 - 9	21	7	147
10 - 14	12	12	144
15 - 19	11	17	187
20 - 24	10	22	220
25 - 29	8	27	216
30 - 34	2	32	64
	80		1010

$$\bar{x} = \frac{1010}{80} = 12.625 \text{ years of teaching experience.}$$

Exercise 3

Find the mean for the grades in a college algebra class.

Class limits	f
35 - 45	2
46 - 56	3
57 - 67	4
68 - 78	8
79 - 89	5
90 -100	2

Example 4--Finding the median.

Find the median cost of speeding tickets for the following sample:
$25 $30 $65 $70 $40 $55 $60 $35 $50 $70 $95

Solution

Put the numbers in order. Find the middle number.
25   30   35   40   50   <u>55</u>   60   65   70   70   95
MD = 55 Half of the speeding tickets were less than $55, half were more than $55.

Exercise 4

Find the median cost of a three item pizza for these sample prices. $7.95 $9.98 $5.58
$4.99 $10.75 $6.25 $8.88 $7.63 $8.50

Example 5--Finding the median.

Find the median time spent on a computer per day for this sample of office workers.
2   6   3   4   2   1   2   0   1   3   6   3

Solution

Put the numbers in order. Since there is an even number of values, the median will be the average of the middle two numbers.
0   1   1   2   2   <u>2   3</u>   3   3   4   6   6

$Md = \dfrac{2+3}{2} = \dfrac{5}{2} = 2.5$ Half of the office workers spent less than 2.5 hours per day on a

computer, half spent more than 2.5 hours on a computer.

Exercise 5

Find the median number of cups of coffee drunk per day by these sample smokers.
6   3   2   4   7   8   6   9   3   5

Example 6--Finding the median for an ungrouped frequency distribution.

Find the median amount spent on a ticket for this concert.

Price (x)	Number of tickets (f)
10	400
15	300
20	450
25	200
50	50

Solution

Find the cumulative frequencies. Look down the *cf* column until you find the first number
that is $\frac{n}{2}$ or larger. $n = 1400$, the total number of frequencies, $\frac{n}{2} = 700$

Price (x)	Number of tickets (f)	cf
10	400	400
15	300	700∗ median class
20	450	1150
25	200	1350
50	50	1400
	1400	

MD = 15, the class value for the median class. Half of the tickets were less than $15.

Exercise 6

Find the median age of children in a fourth grade class.

Ages	Number of students
8	2
9	10
10	8
11	5

Example 7--Finding the median for a grouped frequency distribution.

Find the median number of years of teaching experience for the sample of 80 college professors.

Class limits	f
0 - 4	16
5 - 9	21
10 - 14	12
15 - 19	11
20 - 24	10
25 - 29	8
30 - 34	2

Solution

Find the cumulative frequencies and look down the *cf* column until you find the first number that is $\frac{n}{2}$ or larger. $\frac{n}{2} = \frac{80}{2} = 40$

Class limits	f	cf
0 - 4	16	16
5 - 9	21	37
10 - 14	12	49* median class
15 - 19	11	60
20 - 24	10	70
25 - 29	8	78
30 - 34	2	80
	80	

$$Md = \frac{\frac{n}{2} - cf}{f}(w) + L_m$$

Cf is the cumulative frequency of the class above the median class, $cf = 37$. *F* is the frequency of the median class, $f = 12$. *W* is the width of the median class, $w = 5$. L_m is the lower boundary of the median class, $L_m = 9.5$. (Since the distance from each upper limit to the next lower limit is 1, subtract .5 from the lower class limit to get the lower boundary.)

$$Md = \frac{40 - 37}{12}(5) + 9.5 = \frac{3}{12}(5) + 9.5 = 1.25 + 9.5 = 10.75$$

Half of the professors had less than 10.75 years of teaching experience.

Exercise 7

Find the median for the grades in a college algebra class.

Class limits	f
35 – 45	2
46 – 56	3
57 – 67	4
68 – 78	8
79 – 89	5
90 –100	2

Example 8--Finding the mode.

Find the mode for the cost of speeding tickets for the following sample:
$25 $30 $65 $70 $40 $55 $60 $35 $50 $70 $95

Solution

The mode is the number that appears the most often. $70 is the only number that appears twice, so the mode is $70. There may not be a mode for every set of numbers.

Exercise 8

Find the mode cost of a three-item pizza for these sample prices. $7.95 $9.98 $5.88
$4.99 $10.75 $6.25 $8.88 $7.63 $8.50

Example 9--Finding the mode.

Find the mode time spent on a computer per day for this sample of office workers.
2   6   3   4   2   1   2   0   1   3   6   3

Solution

It easier to see the mode if you put the numbers in order.
0   1   1   2   2   2   3   3   3   4   6   6
Since 2 and 3 both appear three times, they are both modes. When there are 2 modes, this set is called bimodal.

Exercise 9

Find the mode for the number of cups of coffee consumed per day by these sample smokers.
6   3   2   4   7   8   6   9   3   5

Example 10--Finding the mode for an ungrouped frequency distribution.

Find the modal amount spent on a ticket for this concert.

Price (x)	Number of tickets (f)
10	400* modal class
15	300
20	450
25	200

50   50
Solution

The frequency 400 means that there were 400 tickets sold at $10. Since this is the largest frequency, $10 is the mode.

Exercise 10

Find the modal age of children in a fourth grade class.

Ages	Number of students
8	2
9	10
10	8
11	5

Example 11--Finding the mode for a grouped frequency distribution.

Find the mode years of teaching experience for the sample of 80 college professors.

Class limits	f
0 - 4	16
5 - 9	21
10 - 14	12
15 - 19	11
20 - 24	10
25 - 29	8
30 - 34	2

Solution

Find the midpoints by adding each lower and upper limit and dividing by 2.

Class limits	f		x
0 - 4	16		2
5 - 9	21* modal class		7
10 - 14	12		12
15 - 19	11		17
20 - 24	10		22
25 - 29	8		27
30 - 34	2		32

The class with the most frequencies is the modal class. The midpoint of the modal class is the mode. Mode = 7

Exercise 11

Find the mode for the grades in a college algebra class.

Class limits	f
35 - 45	2
46 - 56	3
57 - 67	4
68 - 78	8
79 - 89	5
90 -100	2

Example 12--Finding the midrange.

Find the midrange for the costs of this sample of speeding tickets. $25 $30 $65 $70

$40 $55 $60 $35 $50 $70 $95
Solution

$$\text{midrange} = \frac{\text{highest value} + \text{lowest value}}{2} = \frac{95 + 25}{2} = \frac{120}{2} = 60$$

Exercise 12

Find the midrange for the number of cups of coffee consumed per day by these smokers.
6   3   2   4   7   8   6   9   2   5

Example 13--Finding the range.

Find the range for the costs of this sample of speeding tickets. $25 $30 $65 $70
$40 $55 $60 $35 $50 $70 $95

Solution

$$\text{range} = \text{highest value} - \text{lowest value} = 95 - 25 = \$70$$

Exercise 13

Find the range for the number of cups of coffee consumed per day by these sample smokers.
6   3   2   4   7   8   6   9   2   5

Example 14--Finding the standard deviation.

Find the standard deviation for the cost of speeding tickets for the following sample:
$25 $30 $65 $70 $40 $55 $60 $35 $50 $70 $95

Solution

$$s = \sqrt{\frac{\Sigma x^2 - \frac{(\Sigma x)^2}{n}}{n-1}}$$
You need to add up all the values and add up the square of each value. It is easier to do this work in columns.

x	x^2
25	625
30	900
65	4225
70	4900
40	1600
55	3025
60	3600
35	1225
50	2500
70	4900
95	9025
595	36,525

$n = 11$ since there are 11 numbers in the set

$$s = \sqrt{\frac{36,525 - \frac{595^2}{11}}{10}} = \sqrt{434.09} = 20.83$$

The numbers are deviating about 21 from the mean. This is a fairly large deviation.

Exercise 14

Find the standard deviation for the number of cups of coffee consumed per day by these
sample smokers.   6   3   2   4   7   8   6   9   3   5

Example 15--Finding the standard deviation for a grouped frequency distribution.

Find the standard deviation for the years teaching experience for the sample of 80 college
professors.

Class limits	f
0 - 4	16
5 - 9	21
10 - 14	12
15 - 19	11
20 - 24	10
25 - 29	8
30 - 34	2

Solution

$$s = \sqrt{\frac{\sum fx^2 - \frac{(\sum fx)^2}{n}}{n - 1}}$$

You need the midpoints, so add the lower and upper limits for each class and divide by 2.
Then make a column that is each frequency multiplied by its midpoint (fx). Add those
products. Then you need a column that is fx^2, so multiply each number in the fx column by
x since $(fx)x = fx^2$.

Class limits	f	x	fx	fx^2
0 - 4	16	2	32	64
5 - 9	21	7	147	1029
10 - 14	12	12	144	1728
15 - 19	11	17	187	3179
20 - 24	10	22	220	4840
25 - 29	8	27	216	5832
30 - 34	2	32	64	2048
	80		1010	18,720

$$s = \sqrt{\frac{18,720 - \frac{1010^2}{80}}{79}} = \sqrt{75.5537947} = 8.69$$

To do this work on a calculator, punch in the numbers and operations in the numerator just
as you see them in the formula and then hit the equal button. Then divide by 79 and hit
the equal button again. Last, do the square root. To tell if this is a large deviation or not,
you must compare the standard deviation to the mean. The mean for this problem was
found in example 3 to be 12.625. For a mean of 12.625, 8.69 is a large deviation. The years
of teaching experience for this sample are not close together, the numbers are fairly far
apart. To make it easier to see if the standard deviation is large or not; find the coefficient
of variation.

$CV = \frac{s}{\bar{x}}(100\%) = \frac{8.69}{12.625}(100\%) = 68.83\%$ If CV is more than 50%, there is a large
deviation.

Exercise 15

Find the standard deviation for the grades in a college algebra class.

Class limits	f
35 - 45	2
46 - 56	3
57 - 67	4
68 - 78	8
79 - 89	5
90 -100	2

Example 16--Using the z score.

Which score has the higher relative position: a score of 68 on a test with $\bar{x} = 75$ and $s = 5.6$ or a score of 60 on a test with $\bar{x} = 72$ and $s = 7$?

Solution

Use z scores to compare values from data sets with different means. $z = \frac{x - \bar{x}}{s}$ where x is the given score.

For the first score, $z = \frac{68 - 75}{5.6} = \frac{-7}{5.6} = -1.25$. For the second score, $z = \frac{60 - 72}{7} = \frac{-12}{7} = -1.71$.

Since we want the higher relative score, -1.25 is higher. The first score is better.

Exercise 16

Which score has the higher relative position: a score of 560 on a test with $\bar{x} = 480$ and $s = 45$ or a score of 20 on a test with $\bar{x} = 16$ and $s = 3$?

Example 17--Using Chebyshev's Theorem.

For a data set with mean 212 and standard deviation 18, what percent of the values will fall between 185 and 239?

Solution

To find k, first find how far from the mean is each boundary. $212 - 185 = 27$
$239 - 212 = 27$
Find out how many standard deviations 27 represents.

$\frac{27}{18} = 1.5 = k$ $\qquad 1 - \frac{1}{k^2} = 1 - \frac{1}{1.5} = .556$

55.6% of the values fall between 185 and 239.

Exercise 17

For a data set with $\mu = 1850$ and $\sigma = 72$, what percent of the values will fall between 1698.8 and 2001.2?

Example 18--Using Chebyshev's Theorem.

80% of the values will fall between what two numbers for a data set with $\mu = 105$ and $\sigma = 12$?

Solution

Find k for 80% of the values. (80% = .8)

$1 - \dfrac{1}{k^2} = .8$

$-\dfrac{1}{k^2} = -.2$

$\dfrac{1}{k^2} = .2$

$k = 2.23 \qquad \qquad \begin{aligned} \mu + k\sigma &= 105 + 2.23(12) = 131.76 \\ \mu - k\sigma &= 105 - 2.23(12) = 78.24 \end{aligned}$

80% of the values will fall between 78.24 and 131.76.

Exercise 18

For $\mu = 2376$ and $\sigma = 172$, 90% of the data will fall between what two numbers?

Example 19--Using the empirical rule.

In a normally distributed group of numbers, 95% of the data will fall between what two values if $\mu = 18$ and $\sigma = 5$?

Solution

We can use the empirical rule since we know the data is normally distributed. 95% of the values will fall within 2 standard deviations of the mean.
$$\mu + 2\sigma = 18 + 2(5) = 28 \qquad\qquad \mu - 2\sigma = 18 - 2(5) = 8$$
95% of the values will fall between 8 and 28.

Exercise 19

In a normally distributed group of numbers, 99.7% of the data will fall between what two values if $\mu = 18$ and $\sigma = 5$?

Example 20--Finding percentiles.

Find p_{25}, p_{60}, and the percentile for the time spent on a computer for these office workers.

2   6   3   4   2   1   2   0   1   3   6   3

Solution

To find p_{25}:
1. Order the numbers: 0   1   1   2   2   2   3   3   3   4   6   6
2. Find C. $C = \dfrac{n(\text{percentile})}{100} = \dfrac{12(25)}{100} = 3$
3. Since C is a whole number, average the numbers in the 3rd and 4th positions.

$$p_{25} = \frac{1+2}{2} = \frac{3}{2} = 1.5 \qquad 25\% \text{ of the values are less than } 1.5.$$

To find p_{60}:
1. Order the numbers: 0 1 1 2 2 2 3 3 3 4 6 6

2. Find C. $C = \dfrac{n(\text{percentile})}{100} = \dfrac{12(60)}{100} = 7.2$

3. Since 7.2 is not a whole number, round up to 8. Find the number in the 8th position.
    $p_{60} = 3 \qquad 80\%$ of the values are less than 3.

To find the percentile for 4:

$$\text{percentile} = \frac{\text{number of values below } 4 + .5}{\text{total number of values}}(100\%) = \frac{9.5}{12}(100\%) = 79\%$$

Exercise 20

Find p_{40} and the percentile for 7 for the number of cups of coffee consumed per day.
6   3   2   4   7   8   6   9   3   5

Example 21--Finding percentiles for a frequency distribution.

Find the percentile using the frequency distribution from Example 3 for a teacher with 28 years of experience and find p_{37}.

Solution

Draw a relative frequency ogive.

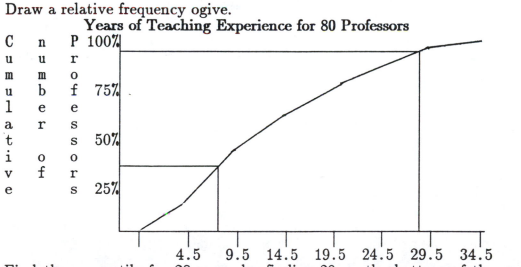

Find the percentile for 28 years by finding 28 on the bottom of the graph and moving up until you hit the line. Then move across until you can read the percentile. The percentile for 28 years is approximately 96. 96% of the teachers had less than 28 years of teaching experience. Find the score for 37% by starting at the side and them moving down and reading that the number is about. 37% of the teachers had less than 8 years teaching experience.

Exercise 21

Find p_{20} and the percentile for a student with a grade of 85 in this college algebra class.

Class limits	f
35 - 45	2
46 - 56	3
57 - 67	4
68 - 78	8
79 - 89	5
90 -100	2

Example 22--Making a box and whisker plot.

Make a box and whisker plot for the number of hours spent on a computer for this sample of office workers.

2   6   3   4   2   1   2   0   1   3   6   3

Solution

Put the values in order.   0   1   1   2   2   2   3   3   3   4   6   6
1. highest value = 6
2. lowest value = 0

3. $Md = \dfrac{2+3}{2} = \dfrac{5}{2} = 2.5$

4. Lower hinge is the midpoint of the numbers below the median.
    0   1   1   2   2   2

    lower hinge $= \dfrac{1+2}{2} = \dfrac{3}{2} = 1.5$

5. Upper hinge is the midpoint of the numbers above the median.
    3   3   3   4   6   6

    upper hinge $= \dfrac{3+4}{2} = \dfrac{7}{2} = 3.5$

The values are positively skewed since the right whisker is longer.

Exercise 22

Make a box and whisker plot for the following number of cups of coffee consumed per day by these smokers.   6   3   2   4   7   8   6   9   2   5

Practice Test

1. Find the mean, the median, the mode, and the midrange. Interpret.
    Number of pages per article in a random sample of magazine articles.
    4   6   4   6   5   4   4   5   7   4   3   6   5   8   4

2. Find the mean, median, and mode. Interpret
    Days of sick leave taken in a month by employees of Franklin Factory

Days of sick leave	Number of employees absent
0	60
1	28
2	5
3	4
4	3

3. Find the mean, median, and mode. Interpret.
    Average number of miles driven per day for 30 commuters

Miles Class limits	Number of commuters f
0 - 24	4
25 - 49	10
50 - 74	11
75 - 99	5

4. Find the standard deviation and interpret.
    Number of pages per article in a random sample of magazine articles.
    4   6   4   6   5   4   4   5   7   4   3   6   5   8   4

5. Find the standard deviation and interpret.

Miles Class limits	Number of commuters f
0 - 24	4
25 - 49	10
50 - 74	11
75 - 99	5

6. A certain pill is supposed to have 50 mg of medicine. A random sample of pills made in plant A was tested and found to have a mean of 47.3 mg of medicine with a standard deviation of 2.1 mg. A sample of pills made in plant B was tested and found to have a mean of 48.3 mg with a standard deviation of 8.2 mg. Use the coefficient of variation to compare the pills made at the two plants.

7. A shopper went to a sale and bought a VCR for $242. The mean price for this VCR at all the stores in the area was $280 with a standard deviation of $26. Another shopper bought a color television for $486 that has a mean price of $548 with a standard deviation of $46. Use the z score to compare these two sales.

8. Use Chebyshev's Theorem to find what percent of the values will fall between 120 and 150 for a data set with mean of 135 and standard deviation of 7.5.

9. Use the Empirical Rule to find what two values 95% of the data will fall between for a data set with mean of 234 and standard deviation of 12.

10. Make a cumulative percent graph to answer the following questions.

Miles Class limits	Number of commuters f
0 - 24	4
25 - 49	10
50 - 74	11
75 - 99	5

a. P_{90}

b. P_{20}

c. Find the percentile for 60 miles.

11. The ages that 25 randomly selected smokers started smoking were:

    26 26 25 17 16 16 14 17 21 15 15 19 16
    17 22 15 19 17 16 16 18 17 16 23 16

    a. Find the score for p_{15}.

    b. Find the percentile for an age of 21.

12. Make a box and whisker plot for the ages that 25 randomly selected smokers started smoking.

    26 26 25 17 16 16 14 17 21 15 15 19 16
    17 22 15 19 17 16 16 18 17 16 23 16

CHAPTER 4
COUNTING TECHNIQUES
Understanding Counting Techniques

Counting techniques are used to find how many different ways there are to do a sequence of events, how many different arrangements of items, or how many different combinations of items are possible.

Multiplication rules are used to find the possible number of ways to do a sequence of events. Multiply the number of ways to do each event together to get the total number of ways to do all the events.

Multiplication Rule - If there are m ways to do one thing and n ways to do a second thing, there are $(m)(n)$ ways to do both things.

If a person has a choice of 4 shirts, 3 pairs of pants, and 2 pairs of shoes, there are $(4)(3)(2)$ = 24 different ways to get dressed.

Permutations are the possible number of arrangements of objects. Arrangement means that the order of the objects matters. If you want to know the possible number of license plates that can be made using 3 letters followed by 3 numbers, it would be a permutation problem since changing the order of the numbers or letters would result in a different license plate. To find the number of possible permutations, you would multiply the number of choices you have for picking the first object times the number of choices for each succeeding object. The number of license plates possible made with 3 letters followed by 3 numbers would be $(26)(26)(26)(10)(10)(10)$ = 17,576,000 since there are 26 letters to choose from and 10 digits (any digit 0 - 9 could be used).

The number of possible ways to sit 3 people in a row would be $(3)(2)(1)$ = 6 since you have 3 choices of people to put in the first chair, but after the first person is sitting down, you only have 2 choices for the second chair and only 1 choice for the last chair. In the license plate problem, repetition is allowed since you may repeat a number or a letter, but in the second problem, repetition is not allowed since one person cannot sit in two different chairs. When repetition is not allowed, a formula can be used for finding permutations. $_nP_r$ is the possible number of permutation of n objects taken r at a time.

$$_nP_r = \frac{n!}{(n-r)!} \quad \text{when repetitions are not allowed.}$$

"!" is called factorial and means to multiply the number in front of the ! by every whole number less than it until you get to 1. $6! = (6)(5)(4)(3)(2)(1) = 720$. The permutation of 6 objects taken 4 at a time if repetition is not allowed would be:

$$_6P_4 = \frac{6!}{(6-4)!} = \frac{(6)(5)(4)(3)(2)(1)}{(2)(1)} = 360.$$

The same result could be arrived by multiplying the number of choices you have for each pick. When you pick the first one, you have 6 choices then since repetition is not allowed you would have 5 choices, then 4, then 3 or $(6)(5)(4)(3) = 360$.

Another formula for permutations applies for arranging groups of objects where some of the objects are alike.

Permutation of n objects where k_1 are alike, k_2 are alike, etc.:

$$\frac{n!}{k_1!k_2!k_3! \cdots k_j!} \quad \text{where } n \text{ is the total number of objects and } j \text{ is the number of groups you have.}$$

Combinations are the possible groups of objects. You are looking for groups when the order does not matter. The number of combinations of r objects taken from a total of n objects is $_nC_r$.

$$_nC_r = \frac{n!}{(n-r)!\,r!}$$

Tree diagrams are used to show all possibilities of a sequence of events. If a family has two children the following tree diagram shows all the possible outcomes.

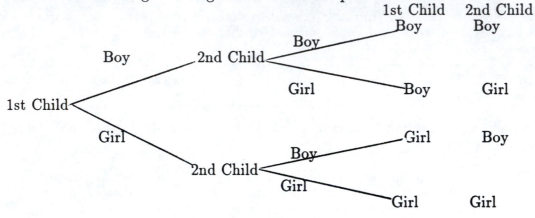

NOTES:

Checking Your Understanding

Complete this section before you do the exercises to make sure you understand the concepts. Write in the book. The answers are in the back of the book. Make a note of any questions that you wish to discuss with the instructor.

I. Tell if each of the following is a combination or permutation and explain why.
   a. The possible batting lineup for nine men on a baseball team.

   b. There are 10 people and 3 are chosen for a committee.

   c. Five girls are picked from a total of 8 to be the starting basketball team.

   d. First, second, and third place prizes are awarded to three of five contestants.

   e. Five scholarships of $1000 each are to be given to 5 students from a group of 15.

   f. The 8 football teams in a conference must play each other team once.

II. Select the correct answer and write the appropriate letter in the space provided.

_____1. If the order matters, then you are looking for the number of
      a. combinations.
      b. permutations.
      c. tree diagrams.

_____2. If the order does not matter, then you are looking for the number of
      a. combinations.
      b. permutations.
      c. tree diagrams.

_____3. The number of groups is the number of
      a. combinations.
      b. permutations.
      c. tree diagrams.

_____4. The number of arrangements is the number of
    a. combinations.
    b. permutations.
    c. tree diagrams.

QUESTIONS:

Applying Your Understanding

STUDENT: In the preceding sections, you learned the concepts of counting techniques, and checked your understanding of the concepts. In this section, you will apply your understanding of the concepts. Study each example carefully and then try to work the following exercise. If you have any problems, see your instructor. The answers to the exercises are in the back of the book.

Example 1--Using the multiplication rule.

If data is grouped by sex (male or female) and by age group (0 - 30, 31 - 50, 51 - 65, 66 and over), how many different categories are possible?

Solution

Draw a box or blank line to represent how many choices you have to make. Fill in each blank with the number of choices you have and multiply them together.

$\underline{2}$ $\underline{4}$ $= (2)(4) = 8$
sex age

> **Exercise 1**
>
> A student has a choice of 3 sections of English, 4 sections of mathematics, 2 sections of science, and 5 sections of history. How many different schedules are possible if the student takes one section of these courses?

Example 2--Using the multiplication rule.

How many different baseball batting lineups are possible for nine players if the coach wants to put one of the two best batters in the third position and wants the pitcher in the last position?

Solution

Draw nine blanks for each of the nine batting positions. Look at the special conditions first. Since there are only 2 choices for the third position and only one choice for the last position, fill in 2 in the third blank and 1 in the last blank.

$\underline{\hphantom{7}}$ $\underline{\hphantom{6}}$ $\underline{2}$ $\underline{\hphantom{5}}$ $\underline{\hphantom{4}}$ $\underline{\hphantom{3}}$ $\underline{\hphantom{2}}$ $\underline{\hphantom{1}}$ $\underline{1}$

Since two players are already assigned, there are now 7 players left. There are 7 choices for the first position, 6 for the second, 5 for the fourth and so on.

$\underline{7}$ $\underline{6}$ $\underline{2}$ $\underline{5}$ $\underline{4}$ $\underline{3}$ $\underline{2}$ $\underline{1}$ $\underline{1}$ $= (7)(6)(2)(5)(4)(3)(2)(1)(1) = 10{,}080$

There are 10,080 different lineups using the coach's requirements.

> **Exercise 2**
>
> Five girls on a basketball team are going to get uniforms numbered 1, 2, 3, 4, and 5. How many ways can the uniforms be given if Wanda wants to be #2 since that was her number last year?

Example 3--Permutations.

A manager wants to visit 6 different cities where the company has branch offices.   How many different ways are there to visit the 6 cities?

Solution

This is a permutation problem because you are looking for possible arrangements of the cities (order matters).   Repetitions are not allowed since you would not go to the same city twice.   It is a permutation of 6 cities taken 6 at a time so $n = 6$ and $r = 6$.

$$_nP_r = \frac{n!}{(n - r)!} = \frac{6!}{(6 - 6)!} = \frac{6!}{0!} = \frac{6!}{1} = 720 \qquad \text{(0! is defined to be 1.)}$$

Permutation problems can also be worked by drawing blanks for each choice and filling the blanks with the number of choices and multiplying the numbers together.

$\underline{6}$ $\underline{5}$ $\underline{4}$ $\underline{3}$ $\underline{2}$ $\underline{1}$ $= (6)(5)(4)(3)(2)(1) = 720$ possible ways to arrange the cities.

> **Exercise 3**
>
> Three students are going to give a report in class. How many different ways can they be arranged?

Example 4--Permutations.

How many ways can 1st, 2nd, and 3rd place be awarded to 7 contestants?

Solution

Order matters, so this is a permutation.   Repetitions are not allowed since you would not award two awards to the same person, so the formula can   be used or you can fill in the blanks.

$\underline{7}$ $\underline{6}$ $\underline{5}$ $= (7)(6)(5) = 210$ ways to award the prize.

> **Exercise 4**
>
> Three children in a class of 17 are going to picked to be awarded prizes for a writing contest. How many possible ways are there to pick the winners if each prize is different?

Example 5--Permutations.

How many different arrangements can be made of the letters in the word bookkeeper?

Solution

Order matters since you are looking for arrangements.   There are groups of like letters, so the second formula is used.

$\frac{n!}{k_1!k_2!k_3! \cdots k_j!}$ n is the total number of objects and the k's are how many are in each group.

There are 10 letters in the word bookkeeper. $n = 10$ There are 1 b, 1 p, 1 r, 2 o's, 2 k's, and 3 e's.

$$\frac{10!}{1!1!1!2!2!3!} = \frac{(10)(9)(8)(7)(6)(5)(4)(3)(2)(1)}{(1)(1)(1)(2)(1)(2)(1)(3)(2)(1)} = 151{,}200$$

Exercise 5

How many different arrangements are there for the letters aaabbbbccddddd?

Example 6--Combinations.

How many ways can 5 scholarships of $1000 each be given to 5 students from a group of 15 applicants?

Solution

This is a combination since the order does not matter. A student picked first gets the same amount as the student picked second. If order does not matter in a counting problem, the combination formula must be used.

$$_nC_r = \frac{n!}{(n-r)!r!} \qquad n \text{ is the total number of objects and } r \text{ is how many to be picked.}$$

$$_{15}C_5 = \frac{15!}{(15-5)!5!} = \frac{(15)(14)(13)(12)(11)(10)(9)(8)(7)(6)(5)(4)(3)(2)(1)}{(10)(9)(8)(7)(6)(5)(4)(3)(2)(1)(5)(4)(3)(2)(1)} = 3003$$

Exercise 6

How many ways can 3 people be picked from 10 to be on a committee?

Example 7--Combinations.

A bowl contains 6 black and 5 white marbles. How many ways can 2 black and 3 white marbles be drawn out?

Solution

You find the ways to pick 2 of the 6 black marbles and the ways to pick 3 of the 5 white marbles out and multiply the answers together. Use the combination formula since the order of the two black marbles does not make a difference since you can not tell the difference between the black marbles.

$$_6C_2 \cdot {}_5C_3 = \frac{6!}{(6-2)!2!} \cdot \frac{5!}{(5-3)!3!} = \frac{(6)(5)(4)(3)(2)(1)}{(4)(3)(2)(1)(2)(1)} \cdot \frac{(5)(4)(3)(2)(1)}{(2)(1)(3)(2)(1)} = 150$$

Exercise 7

How many ways can 3 fifth graders be chosen from 8 fifth graders and 2 fourth graders be chosen from 5 fourth graders?

Example 8--Tree diagram.

Make a tree diagram showing all the possible sandwiches that can be ordered if you can choose from 2 types of rolls, white or rye; 3 kinds of meat, beef, chicken, or turkey; and 2 different cheeses, American and Swiss.

Solution:

Start by drawing branching line for the 2 choices of rolls, from each of branch of the rolls, add 3 branches for the 3 types of meat, and from each of those branches, draw two branches for the 2 types of cheese.

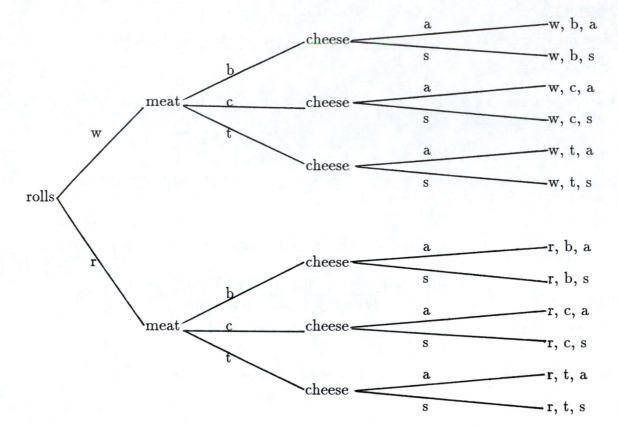

NOTES:

Practice Test

1. There are 10 true-false questions on a quiz. How many different possible keys are there?

2. Four men and 3 women are going to sit in a row for a picture. How many ways can they be placed if a man must sit at each end?

3. A high school principal plans to visit the classrooms of 5 teachers in one day. If there are 12 teachers in the school, how many different ways can the visits be arranged?

4. How many different 3-letter words are possible?

5. How many different arrangements can be made from the letters in the word different?

6. Eight football teams in a conference must play each other team once. How many games must be played?

7. A jury of 12 people is to be chosen from 10 men and 12 women. How many ways can a jury of 6 men and 6 women be chosen?

8. A bowl contains 6 red, 4 black, and 3 yellow marbles. How many ways can 2 red, 2 black, and 1 yellow marble be drawn?

CHAPTER 5
PROBABILITY
Understanding Probability

Introduction

Probability is a measure of the likelihood that a certain outcome will occur in a given experiment. Probabilities can be expressed as a fraction, a decimal, or a percent, but probabilities are always a number between 0 and 1.00 inclusive. An **event** is a set of possible outcomes for an experiment. The **sample space** for an experiment is the set of all possible outcomes for an experiment. There are three types of probabilities; classical, empirical, or subjective.

Classical probability is applied when you can assume each event is equally likely to occur. When you flip a coin, a head or a tail is equally likely to come up. If you have six marbles in a bowl and draw one out without looking, any one of the six marbles has the same chance of being drawn. To find the probability of a particular event in classical probability, you divide the number of ways the event can occur by the total number of outcomes for the experiment. The probability for a particular event E is labeled P(E).

$$P(E) = \frac{\text{number of ways E can occur}}{\text{total number of outcomes in the sample space}}$$

Empirical probability is used when you can not assume that each event is equally likely to occur. You must determine how often like events occurred in the past. Empirical probability is sometimes referred to as relative frequency because you are looking for the frequency of one event in relation to the total number of frequencies. If you want to know the probability of Candidate A winning an election, you would have to ask some voters how they would vote. The probability that the next voter you asked would vote for Candidate A would be the number that said they would vote for A divided by the total number of voters interviewed.

$$P(E) = \frac{\text{number in event E in the past}}{\text{total number of outcomes in the sample space}}$$

Subjective probability is used when each event can not be assumed to be equally likely and you can not perform the experiment to determine the relative frequencies. In subjective probabilities, an informed person makes an educated guess. When the weatherman gives the probability of rain tomorrow, he is giving his guess. Subjective probability is only as good as the knowledge the person has who is making the probability.

Special Definitions

Mutually exclusive events are two events that cannot occur at the same time. If you flip a coin you can not get a head and a tail on the same toss, so the events "head" and "tail" are mutually exclusive. If you roll a die (singular of the word dice, assumed to be a six-sided die), the event that you get an even number and the event that you get a number less than 4 are not mutually exclusive because the number two appears in both events.

Two events are **independent events** if the outcome of the first event does not effect the probability of the second event. If you toss two dice, the outcome on the first die has no effect on the outcome of the second die. If the outcome of the first event affects the probability of the second event, the events are **dependent events**. If you draw two marbles from a bowl that has ten red and five blue marbles without replacing the first one drawn, the probability of the second draw depends on whether you got a blue or a red marble on the first draw.

The **complement** of an event is all the outcomes not in that event. The probability of the complement is the probability that the event will not occur. The complement of E is denoted by $\overline{E}$.

Rules of Probability

There are rules of probability that help determine probabilities. The first five rules apply to all probability problems.

Rule 1: $0 \leq P(E) \leq 1.00$.

Rule 2: $P(E) = 0$ means E cannot occur.

Rule 3: $P(E) = 1$ means E will occur.

Rule 4: $P(\overline{E}) = 1 - P(E)$ and $P(E) = 1 - P(\overline{E})$.

Rule 5: **The sum of the probabilities in the sample space is 1.**

The **addition rules of probability** are applied when you are looking for the probability of one event or another event occurring. The key word "or" tells us to add probabilities. There are two addition rules.

The first rule of addition applies if the two events are mutually exclusive. To find the probability of one event or another when they are mutually exclusive, you add the probability of the first event to the probability of the second event. To find the probability of getting a 2 or a 3 when you roll a die, you add the probability of getting a 2 to the probability of getting a 3.

$$P(2 \text{ or } 3) = P(2) + P(3) = \frac{1}{6} + \frac{1}{6} = \frac{2}{6} = \frac{1}{3} \text{ or } .333.$$

$$\boxed{P(A \text{ or } B) = P(A) + P(B) \text{ if A and B are mutually exclusive.}}$$

The second rule of addition applies if the events are not mutually exclusive. To find the probability of one event or another event if they are not mutually exclusive, you add the probability of the first event to the probability of the second event and subtract the probability of anything that was common to both events. To find the probability of getting a king or a heart when you draw a card from a standard deck of fifty two cards, you add the probability of getting a king to the probability of getting a heart and subtract the probability of getting a king of hearts. (You may want to review the suits and number of cards in a deck since this type of problem appears frequently in probability.)

$$P(\text{king or heart}) = P(\text{king}) + P(\text{heart}) - P(\text{king of hearts}) = \frac{4}{52} + \frac{13}{52} - \frac{1}{52} = \frac{16}{52} = \frac{4}{13}$$

$$\boxed{P(A \text{ or } B) = P(A) + P(B) - P(A \text{ and } B) \text{ if A and B are not mutually exclusive.}}$$

The **multiplication rules of probability** apply when you are looking for the probability of one event and another event. The key word "and" tells you to multiply probabilities. The probability of one event and another event is called a **joint probability**.

The first rule of multiplication is used when the two events are independent. If A and B are independent, the probability of A and B happening is the probability of A times the probability of B. The probability of rolling two sixes when you roll two dice is the probability of a six on the first roll times the probability of a six on the second roll.

$$P(\text{six and six}) = P(\text{six}) \cdot P(\text{six}) = \frac{1}{6} \cdot \frac{1}{6} = \frac{1}{36}$$

$$\boxed{\textbf{P(A and B)} = \textbf{P(A)} \cdot \textbf{P(B) if A and B are dependent.}}$$

The second rule of probability applies when the two events are not independent. If the two events are dependent, the probability of the first and the second event is the probability of the first event times the probability of the second event assuming the first event already happened. The probability of the second event assuming the first event happened is called the **conditional probability**. The probability of drawing two cards from a standard deck and getting two aces is equal to the probability of getting an ace on the first draw times the probability of getting an ace on the second draw assuming that you have already drawn one ace. The probability of getting a second ace when you assume you have already drawn one ace is the conditional probability and is represented by $P(B/A)$ and is read the probability of B given A.

$$P(\text{two aces}) = P(\text{ace and ace}) = P(\text{ace}) \cdot P(\text{ace/ace}) = \frac{4}{52} \cdot \frac{3}{51} = \frac{12}{2652} = \frac{1}{221}$$

$$\boxed{\textbf{P(A and B)} = \textbf{P(A)P(B/A) if A and B are dependent events.}}$$

Bayes' Theorem is used to find conditional probabilities. The probability of A given B can be found using Bayes' Theorem if the probability of B given A is known. Bayes' Theorem is used if the event A can occur in A_1, A_2, $\cdots A_n$ mutually exclusive ways and the event B can occur in B_1, B_2, $\cdots B_n$ mutually exclusive ways.

$$\boxed{\begin{array}{l} \textbf{Bayes' Theorem} \\[6pt] P(A_1/B_1) = \dfrac{P(A_1) \cdot P(B_1/A_1)}{P(A_1) \cdot P(B_1/A_1) + P(A_2) \cdot P(B_1/A_2) + \cdots + P(A_n)P(B_1/A_n)} \end{array}}$$

The easiest way to apply Bayes' Theorem is to draw a tree diagram. The numerator is the product of the branch that has A_1 and B_1. The denominator is the sum of all the products of the branches that contain B_1.

NOTES:

Checking Your Understanding

Complete this section before you do the exercises to make sure you understand the concepts. Write in the book. The answers are in the back of the book. Make a note of any questions that you wish to discuss with the instructor.

I. Define
   1. Mutually exclusive events

   2. Independent events

   3. Complement

II. Discuss the difference between classical and empirical probabilities.

III. Discuss the two rules of addition. Give similarities and differences.

IV. Discuss the two rules of multiplication. Give similarities and differences.

V. Select the correct answer and write the appropriate letter in the space provided:

_____1. In classical probability, the events are assumed to be
   a. independent.
   b. mutually exclusive.
   c. equally likely.
   d. between -1.00 and 1.00.

_____2. If the outcome of one event does not affect the probability of another event, the two events are
   a. dependent.
   b. mutually exclusive.
   c. independent.
   d. complementary.

_____3. If you want to find the probability of one event or another, which rules would you use?
  a. addition
  b. multiplication

_____4. A collection of all the possible outcomes of an experiment is the
  a. event.
  b. complement.
  c. null set.
  d. sample set.

_____5. If you want to find the probability of one event and another, which rules would you use?
  a. addition
  b. multiplication

_____6. If the event E cannot occur, then P(E) =
  a. −1.00.
  b. 0.
  c. 1.00.
  d. 100.

_____7. If the event E will definitely occur, then P(E) =
  a. −1.00.
  b. 0.
  c. 1.00.
  d. 100.

_____8. The probability of an event happening given that another event has already happened is
  a. an event.
  b. a probability.
  c. a joint probability.
  d. a conditional probability.

_____9. A probability can have a value between
  a. −1.00 and 1.00.
  b. 0 and 100.
  c. 0 and 1.00.
  d. −100 and 100.

_____10. If a sportscaster gives the probability of a football team winning as 30%, he is using
  a. classical probability.
  b. subjective probability.
  c. empirical probability.
  d. relative probability.

_____11. A joint probability is the probability of
  a. one event and another event.
  b. one event or another event.
  c. one event occurring given that another event has occurred.
  d. an event that is certain to happen.

_____12. A conditional probability is the probability of
  a. one event and another event.
  b. one event or another event.
  c. one event occurring given that another event has occurred.
  d. an event that is certain to happen.

Applying Your Understanding

Student: In the preceding sections, you learned the concepts of probability, and checked your understanding of the concepts. In this section, you will apply your understanding of the concepts. Study each example carefully and then try to work the following exercise. If you have any problems, see your instructor. The answers are in the back of the book.

Example 1--Finding classical probability.

There are 8 red, 4 blue, and 6 white marbles in a jar. If you draw one out without looking, find the probability of getting a red marble.

Solution

This is classical probability since you can assume each marble is equally likely to be drawn. Divide the number of red marbles by the total number of marbles.

$$P(\text{red}) = \frac{8}{18} = \frac{4}{9}$$

Exercise 1

Find the probability of getting a blue marble if you draw one marble from a jar of 8 red, 4 blue, and 6 white marbles.

Example 2-- Finding empirical probability.

750 people were interviewed and asked to name the vice president of the United States. 525 people answered correctly. Find the probability that the next person asked can correctly identify the vice president.

Solution

This is empirical probability so divide the number of people who answered correctly by the total number of people interviewed.

$$P(\text{correct}) = \frac{525}{750} = \frac{7}{10}$$

Exercise 2

There are 18 freshmen and 23 sophomores in a class. If one student is chosen at random to answer a question, find the probability that the one chosen is a freshman.

Example 3--Using the first addition rule.

A company has 66 employees. If one employee is chosen at random, what is the probability that a maintenance or a production worker is chosen.

Maintenance   14
Production   37
Office staff   8
Management   7

Solution

The events are mutually exclusive. Add the probabilities.
$$P(\text{maintenance or production}) = P(\text{maintenance}) + P(\text{production}) = \frac{14}{66} + \frac{37}{66} = \frac{51}{66} = \frac{17}{22}$$

Exercise 3

There are members of four political parties at a rally, 67 Democrats, 13 Republicans, 17 Independents, and 12 American Party. Find the probability of picking one to interview and getting either an Independent or a member of the American Party.

Example 4--Using the second rule of addition.

36 people apply for a job, 20 men and 16 women. 8 of the men and 12 of the women have Ph.D.s. If one person is selected at random for an interview, find the probability that the one chosen is a woman or has a Ph.D.

Solution

The events are not mutually exclusive, so you must find the probability of a woman add the probability of a Ph.D. and subtract the probability of a woman and a Ph.D.

$$P(\text{woman or Ph.D.}) = P(\text{woman}) + P(\text{Ph.D.}) - P(\text{woman and Ph.D.}) = \frac{16}{36} + \frac{20}{36} - \frac{12}{36}$$
$$= \frac{24}{36} = \frac{2}{3}$$

Exercise 4

At a party, there are 32 men, 16 of whom are without dates, and 28 women, 12 of whom are without dates. Find the probability of picking someone at random and getting a person who is a woman or without a date.

Example 5--Using the first rule of multiplication.

If 17% of the people in the United States smoke, find the probability of picking three people at random and finding that all three smoke.

Solution

All smoke means the first one smokes, the second one smokes, and the third one smokes, so this is a multiplication problem. These are independent events since whether the first one smokes has no effect on the probability that the second one smokes. A percent can be thought of as a probability, so 17% means the probability of getting a smoker is .17. Multiply the probabilities together.

$$P(\text{all three smoke}) = P(\text{smoke})P(\text{smoke})P(\text{smoke}) = (.17)(.17)(.17) = .004913$$

Exercise 5

44% of the professors at four year colleges are female. If you sign up for five classes, find the probability that you get all female professors.

Example 6--Using the second rule of addition.

If 47% of the people read a newspaper every day, 62% of the people watch the television news, and 25% do both, find the probability of picking a person at random and getting someone who reads the newspaper or watches the news on television.

Solution

Add the probability of reading the newspaper to the probability of watching television news and subtract the probability of both since the events are not mutually exclusive.

P(newspaper or watch television) = P(newspaper) + P(watch television) − P(both)
= .47 + .62 − .25 = .84

Exercise 6

If 38% of the people eat out once a week, 42% take home prepared food once and 29% do both, find the probability of selecting someone who eats out or takes home once a week.

Example 7--Using the second rule of multiplication.

A box of 35 computer components has 6 defectives. If you test 2 of these components, find the probability that both work properly.

Solution

Both work properly means the first one works and the second one works. "And" tells you to multiply. These are not independent events since the probability of the second trial depends on whether you removed a good or a defective component on the first try. (Assume that if you test one component, you would not put it back in the box before testing the second one.) If there are 6 defective parts out of 35, there must be 29 good parts. The probability of a good part on the first try is $\frac{29}{35}$. For the second event, you have to assume that you got the event that you wanted on the first try, so there would be 28 good ones left and 34 total components left. The probability of a good on the second try would be $\frac{28}{34}$.

P(good and good) = P(good)P(good/good on first try) = $\frac{29}{35} \cdot \frac{28}{34}$ = .6824

Exercise 7

In a class of 30 students, 5 made A's on the first test. If 3 students are selected at random to do a special project, find the probability that all three students made A's on the first test.

Example 8--Using the second rule of multiplication.

23% of a class passed the first and second test. 45% of those that passed the first test also passed the second test. Find the percent that passed the first test.

Solution

Since percents can be treated like probabilities, we can find the probability of passing the first test and give the answer as a percent.
P(passing first test and second) = .23
P(passing the second/passed the first) = .45
Use the formula for A and B if the events are dependent.
P(A and B) = P(A)P(B/A)
P(passing first and second) = P(passing first)P(passing the second/passed the first)
.23 = P(passing first)(.45)   Divide both sides of the equation by .45.
.51 = P(passing first)
51% passed the first test.

> **Exercise 8**
>
> The probability of a freshman college student passing English I and English II is .31. The probability of passing English I is .58. Find the probability of passing English II given that the student passed English I.

Example 9--Using counting techniques to find a probability.

A delivery route for a bread truck consists of five different stops. If the five stops are picked in random order, find the probability of getting the most effective route.

Solution

Since there is only 1 route that is the most efficient, divide 1 by the total number of routes possible. To find the total number of routes possible, you must use counting techniques. When you pick the first stop you have 5 choices, when you pick the second stop, you have 4 choices, etc.

Total number of routes possible $= 5 \cdot 4 \cdot 3 \cdot 2 \cdot 1 = 120$

$P(\text{most effective route}) = \frac{1}{120}$

> **Exercise 9**
>
> A network chooses 3 shows to put on Monday night. If they choose the order of the shows at random, what is the probability that you have the best order viewer satisfaction?

Exercise 10--Using the complement rule.

Twelve men and 8 women were called for jury duty. If a jury of six is selected, find the probability there will be at least one woman on the jury.

Solution

At least one woman on the jury means you have to find the probability of getting exactly one woman, or exactly 2 women, or exactly 3 women, etc. and add all the probabilities together. Whenever a problem says **at least one**, it is easier to find the complement. The complement of at least one woman is that there are no women on the jury, or all men. Use the conditional probability since these are dependent.

$P(\text{all men}) = \frac{12}{20} \cdot \frac{11}{19} \cdot \frac{10}{18} \cdot \frac{9}{17} \cdot \frac{8}{16} \cdot \frac{7}{15} = .02384$

$P(\text{at least one woman}) = 1 - P(\text{all men}) = 1 - .02384 = .9762$

> **Exercise 10**
>
> There are 26 inmates in a county jail, 5 are in for violent crimes. If 3 are picked at random for a work detail, find the probability that at least a was in for violent crime.

Exercise 11--Using the complement rule.

As a safety factor in a nuclear plant, three safety switches are installed. Each switch has a 5% chance of failing. What is the probability that at least one switch will work?

Solution

The complement of at least one work is that all fail. Find the probability that all three fail and subtract that from 1. P(fail) = .05.

P(all 3 fail) = P(fail)P(fail)P(fail) = (.05)(.05)(.05) = .000125

P(at least on works) = 1 − .000125 = .999875

Exercise 11

In a recent poll of voters, 27% were displeased with both candidates. If 4 people were picked at random, find the probability that at least one liked one of the candidates.

Exercise 12--Using counting techniques to find a probability.

A farmer has three fields to plant. He has a choice of 8 types of cotton, 6 type of corn, or 3 types of milo to plant. If he picks 3 crops at random, find the probability he will plant 2 fields of cotton and 1 of corn.

Solution

There are a combination of 8 items taken 2 at a time ways to choose two types of cotton from 8 types and there are a combination of 6 items taken 1 at a time to choose 1 type of corn. There are a combination of 17 items taken 3 at a time ways to get a total of 3 types from a total of 17 types.

$$P(2 \text{ cotton and } 1 \text{ corn}) = \frac{{}_8C_2 \, {}_6C_1}{{}_{17}C_3} = \frac{\frac{8!}{6!2!} \cdot \frac{6!}{4!2!}}{\frac{17!}{14!3!}} = \frac{28 \cdot 15}{680} = .618$$

Exercise 12

A teacher has a choice of 16 different difficult problems or 5 very difficult problems. If there is a short quiz with 4 questions, and the problems are chosen at random, what is the probability of having 2 difficult problems and 2 very difficult.

Exercise 13--Using Bayes' Theorem.

A test to determine if a patient has a certain rare disease is 99% effective. The disease is known to occur in 0.6% of the population. If a person is tested positive for the disease, find the probability that the person actually has the disease.

Solution

Let D stand for having the disease, N for not having the disease, + for a positive test, and − for a negative test. You are asked to find the probability that a person has the disease, given a positive test. P(D/+) is what you are trying to find.

99% effective means the test was given to a group of people who were known to have the disease, and the test was positive 99% of the time. P(+/D) = .99. Also the test was given to a group of people that did not have the disease and the test was negative 99% of the time. P(−/D) = .99. (In each case, the test was accurate 99% of the time.)
P(D) = .006   P(N) = .994   Draw a tree diagram.

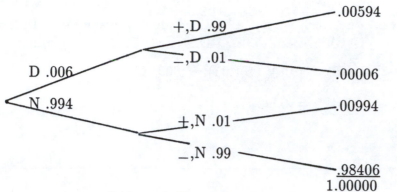

.00594   (multiply .006 times .99)

For the numerator of Bayes' Theorem, find the value at the end of the branch for the probability you want to find, D and +. For the denominator of Bayes' Theorem, add the end of every branch that has the second item that you are trying to find (+).

$$P(D/+) = \frac{.00594}{.00594 + .00994} = .3740554$$

The probability a person has the disease even with a positive test result is only .374.

Exercise 13

Find the probability that a person has the disease if a negative test result is obtained for the information in Example 13.

NOTES:

Practice Test

1. When a card is drawn from a standard deck of 52 cards, what is the probability of getting
   a. a queen?

   b. a red card?

   c. a queen or a red card?

2. A restaurant counted the beverages ordered in one day. 72 people ordered milk, 128 coffee, 86 tea, 32 sodas, and 26 water. If a person is picked at random, what is the probability he orders
   a. coffee?

   b. milk or water?

   c. a drink with caffeine (coffee, tea, or soda)?

3. In a math class of 60 students, 36 were boys and 24 were girls. Five of the boys and 6 of the girls made A's on the first test. If a student is chosen at random, find the probability of getting a boy or an A student.

4. If 67% of all high school graduates go to college, find the probability of selecting 4 high school graduates at random and getting 4 who are going to college.

5. 48% of the adults in America eat health food for breakfast. 36% of the adults exercise regularly. If 26% eat health food for breakfast and exercise, find the probability of an adult eating health food for breakfast or exercising regularly.

6. A shipment of 50 cameras had 6 defectives. If a person bought two cameras, find the probability of getting 2 defectives.

7. The probability that a worker is absent and that it is Monday is .015. Find the probability that a worker is absent given that today is Monday. (Assume a 5 day work week, so the probability of Monday is .2.)

8. At a small company, the employee salaries were compared to their number of years on the job. The data collected is shown in the table.

Salaries	Years 0 - 5	6 - 10	over 10
0 - 19,999	8	3	2
20,000- 39,999	6	5	5
40,000 and up	4	8	3

Find the probability of selecting a employee who has
a. 6 - 10 years on the job.

b. 6 - 10 years on the job or over $40,000 salary.

9. A flu shot is 85% effective in preventing the flu. If 4 people in your family take a flu shot, find the probability that at least one will get the flu.

10. An office has 12 female workers and 6 male workers. If the boss randomly picks 4 workers for a special job, find the probability 2 females and 2 males will be picked.

11. A company orders parts from 2 suppliers, A and B. It gets one box of parts from each supplier. The company knows from past orders that 12% of the parts from A are defective and 20% of the parts from B are defective. One part is selected at random and tested. What is the probability that the part is from supplier B is the part is found to defective?

CHAPTER 6
PROBABILITY DISTRIBUTIONS
Understanding Probability Distributions

Introduction

A **probability distribution** is all of the outcomes of an experiment and the probability for each outcome. There are two rules for a probability distribution. The sum of all the probabilities must be equal to one since a probability distribution includes all the sample space and each probability must be between 0 and 1 inclusive since this is true for all probabilities.

Rules for a Probability Distribution

1. $\sum P(x) = 1$.
2. $0 \leq P(x) \leq 1$.

The following is a probability distribution for the number of boys possible when you have 2 children.

Number of boys x	Probability $P(x)$
0	$\frac{1}{4}$
1	$\frac{1}{2}$
2	$\frac{1}{4}$

The **mean** of a probability distribution is the average number of occurrences you should get if you performed the experiment many times. The mean is sometimes call the **expected value**. The symbol for the mean is μ, and the symbol for the expected value is $E(x)$.

$$\mu = E(x) = \sum x \cdot P(x)$$

The **standard deviation** and **variance** are used to find the variation in the probability distribution. The variance is σ^2 and the standard deviation is σ.

$$\sigma^2 = \sum x^2 P(x) - \mu^2$$
$$\sigma = \sqrt{\sum x^2 P(x) - \mu^2}$$

A **binomial distribution** is a probability distribution that only has two outcomes, called a success and a failure. Flipping a coin can only have two outcomes, a head or a tail. Rolling a die has six possible outcomes, but can be thought of as a binomial if you think of the only two outcomes as a success or failure (getting the number you want or getting any other number). There are four requirements for a binomial probability.

Binomial probabilities have:
1. repeated trials.
2. only two outcomes.
3. independent trials.
4. a constant probability.

Binomial Probability

$$P(x) = \frac{n!}{(n-x)!\, x!}\, p^x q^{n-x}$$

n = total number of trials x = number of successes desired
p = probability of success q = probability of failure = $1 - p$

> **For a binomial distribution:**
>
> | Mean | Variance | Standard deviation |
> |------|----------|--------------------|
> | $\mu = np$ | $\sigma^2 = npq$ | $\sigma = \sqrt{npq}$ |

The **multinomial distribution** is used if there are more than two possible outcomes, the probabilities for each trial remain constant, and the outcomes are independent for a repeated number of trials.

> **Multinomial Probability**
>
> $$P(M) = \frac{n!}{x_1!\, x_2!\, \cdots x_k!}\, p_1{}^{x_1} p_2{}^{x_2} \cdots p_k{}^{x_k}$$
>
> n is the total number of trials
> x_i is the number of times you want each event to occur
> p_i is the probability for each event

The **Poisson distribution** is used when the number of trials (n) is large and the probability (p) is small.

> **Poisson Probability**
>
> $$P(x;\, \lambda) = \frac{e^{-\lambda} \lambda^x}{x!}$$
>
> e = Euler's constant ≈ 2.7183 x = number of occurrences desired
> λ = mean number of occurrences
> or $\lambda = np$ if the probability or percent is given
> or $\lambda = \dfrac{\text{number of desired outcomes in the past}}{\text{total number of outcomes in the past}}$

The **hypergeometric distribution** is used when trials are repeated but the outcomes are not independent so the probability changes for each trial.

> **Hypergeometric Probability**
>
> $$P(A) = \frac{{}_aC_x\, {}_bC_{n-x}}{{}_{(a+b)}C_n}$$
>
> n = total number of items to be picked
> a = number of the first item in the total
> b = number of the second item in the total
> x = number of the first item desired
> $n - x$ = number of the second item desired

NOTES:

Checking Your Understanding

Complete this section before you do the exercises to make sure you understand the concepts. Write in the book. The answers are in the back of the book. Make a note of any questions you wish to discuss with the instructor.

I. Tell if each of the following is a probability distribution. If not, explain.

a.
x	$P(x)$
0	.2
2	.5
4	.1
5	.2

b.
x	$P(x)$
-1	$\frac{1}{2}$
0	$\frac{1}{4}$
1	$\frac{1}{4}$

c.
x	$P(x)$
1	-1
2	0
3	1

d.
x	$P(x)$
2	$\frac{1}{8}$
4	$\frac{1}{8}$
6	$\frac{3}{8}$
8	$\frac{1}{8}$

II. Select the correct answer and write the appropriate letter in the space provided.

_____ 1. If n is large and p is small, use the
  a. binomial distribution.
  b. multinomial distribution.
  c. Poisson distribution.
  d. hypergeometric distribution.

_____ 2. If there are repeated independent trials and there are only two outcomes, use the
  a. binomial distribution.
  b. multinomial distribution.
  c. Poisson distribution.
  d. hypergeometric distribution.

_____ 3. If there are repeated dependent trials, use the
   a. binomial distribution.
   b. multinomial distribution.
   c. Poisson distribution.
   d. hypergeometric distribution.

_____ 4. If there are repeated independent trials with more than two outcomes, use the
   a. binomial distribution.
   b. multinomial distribution.
   c. Poisson distribution.
   d. hypergeometric distribution.

_____ 5. The expected value is the same as the
   a. mean.
   b. variance.
   c. standard deviation.
   d. probability distribution.

QUESTIONS:

Applying Your Understanding

STUDENT: In the preceding sections, you learned the concepts of probability distributions, and checked your understanding of the concepts. In this section, you will apply your understanding of the concepts. Study each example carefully and then try to work the following exercise. If you have any problems, see your instructor. The answers are in the back of the book.

Example 1--Finding the expected value and standard deviation.

A store finds the following results for profits and losses for each day:

Profit or loss, x	$P(x)$
$1000	.4
$-$2000	.3
$-$1000	.2
$5000	.1

Find the expected value and the standard deviation.

Solution

$E(x) = \sum x \cdot P(x)$ and $\sigma = \sqrt{\sum x^2 P(x) - \mu^2}$. Make two columns, $x \cdot P(x)$ and $x^2 P(x)$.

x	$P(x)$	$x \cdot P(x)$	$x^2 P(x)$
1000	.4	400	400,000
$-$2000	.3	$-$600	1,200,000
$-$1000	.2	$-$200	200,000
5000	.1	500	2,500,000
		100	4,300,000

$E(x) = \$100$ $\sigma = \sqrt{4,300,000 - 100^2} = \sqrt{4,290,000} = \$2071$

The expected value is $100. Over a long period of time, the average daily profit should be $100, but $2071 is a large deviation, so the company should prepare for days with large losses and some with large gains.

Exercise 1

Find the expected value and the standard deviation for winning $1000 if 2000 tickets are sold and you buy one ticket for $1. Only one ticket wins.

Example 2--Binomial probabilities.

For a certain company, workers miss work 17% of the days. If one department has 12 employees, find the probability of no absences on a given day and find the probability of 3 absences on a given day.

Solution

This is a binomial problem since there are only two outcomes (absent or present), the trials are independent, and the probability is constant (.17 for all trials).

$P(x) = \dfrac{n!}{(n-x)! \, x!} \, p^x q^{n-x}$ n = total number of trials = 12
$\phantom{P(x) = \dfrac{n!}{(n-x)! \, x!} \, p^x q^{n-x}}$ x = number of successes desired = 0
p = probability of success = .17 q = probability of failure = $1 - p = 1 - .17 = .83$

$$P(0) = \frac{12!}{(12 - 0)!0!}(.17)^0(.83)^{12} = .107 \qquad (0! = 1)$$

$$P(3) = \frac{12!}{(12 - 3)!3!}(.17)^3(.83)^9 = .202$$

Exercise 2

If 27% of the cars in America are some shade of blue and there are 10 cars in a parking lot, find the probability that 5 of the cars will be blue.

Example 3--Binomial tables.

40% of a certain community is Hispanic. If 12 people are in a restaurant, find the probability that less than 4 are Hispanic.

Solution

Less than 4 would be if 0, 1, 2, or 3 were Hispanic. You have to use the formula for $x = 0$, 1, 2, and 3 and then add the probabilities since the word "or" tells us to add. Binomial probabilities can be looked up in Table B if the probability desired is listed on the table. Look at Table B in the back of the text or in the back of this book. Since $p = .4$ is on the table, look at the far left column until you find $n = 12$. Look at the second column until you find the desired x and read the answer.
P(less than 4) = P(0) + P(1) + P(2) + P(3) = .002 + .017 + .064 + .142 = .225

Exercise 3

10% of the cameras sold in the United States are manufactured in Europe. If 20 cameras are selected at random, what is the probability that at least 3 were manufactured in Europe.

Example 4--Mean and standard deviation for a binomial distribution.

If a company has 200 employees and 17% are absent each day, find the mean and standard deviation for the number of absences each day.

Solution

$$\mu = np \qquad \sigma = \sqrt{npq} \qquad\qquad n = 200 \qquad\qquad p = .17 \qquad\qquad q = 1 - .17 = .83$$

$$\mu = 200(.17) = 34 \qquad\qquad \sigma = \sqrt{200(.17)(.83)} = \sqrt{28.22} = 5.3$$

There will be an average of 34 absences a day with a standard deviation of 5.3 absences.

Exercise 4

56% of the people wear seat belts. If 50 people were randomly checked, find the mean and standard deviation for how many will be wearing their seat belts.

Example 5--Multinomial probability.

The probability of an A in a certain course is .1, B is .2, C is .5, and lower than a C is .2. If 20 people are in the class, find the probability of 3 A's, 5 B's, 10 C's, and 2 below C.

Solution

This is a multinomial since there are repeated independent trials but there are more than 2 possible outcomes.

$$P(M) = \frac{n!}{x_1! \, x_2! \cdots x_k!} \, p_1^{x_1} p_2^{x_2} \cdots p_k^{x_k}$$

n is the total number of trials
x_i is the number of times you want each event to occur
p_i is the probability for each event

$n = 20$
$x_1 = 3 \qquad p_1 = .1$ (number of A's desired and the probability of an A)
$x_2 = 5 \qquad p_2 = .2$ (B's)
$x_3 = 10 \qquad p_3 = .5$ (C's)
$x_4 = 2 \qquad p_4 = .2$ (lower than C)

$$P = \frac{20!}{3!5!10!2!}(.1)^3(.2)^5(.5)^{10}(.2)^2 = .00582$$

Exercise 5

Find the probability of 2 A's, 4 B's, 12 C's and 2 below C for the class in Example 5.

Example 6--Poisson.

A door-to-door salesman averages 2 sales a day. Find the probability of getting 5 sales in a day if this approximates a Poisson distribution.

Solution

$P(x; \lambda) = P(5, 2)$ where x is the number desired and λ is the average number of occurrences. On Table C, look across the rows labeled at the top until you find $\lambda = 2$. Then look under the 2 column until you find the row for $x = 5$. $P(5; 2) = .0361$.

Exercise 6

A checkout line at a grocery store has an average of 5 people waiting. What is the probability of 7 people waiting if this approximates a Poisson?

Example 7--Poisson.

There is a probability of .2% of a defective part in a shipment of computer components. If a shipment of 500 components arrives, find the probability of 0 defectives. This approximates a Poisson.

Solution

$\lambda = np$ when a percent or probability is given $\lambda = 500(.002) = 1$ $x = 0$ $P(0; 1) = .3679$

Exercise 7

A company has 1000 office machines with a probability of .005 of breaking down in a day. What is the probability of 6 breakdowns in a day if this approximates a Poisson?

Example 8--Poisson.

1000 cotton bolls were checked for boll weevils and 100 weevils were found. What is the probability of finding a boll with 2 weevils if this approximates a Poisson.

Solution

$$\lambda = \frac{\text{number of desired item in the past}}{\text{total number in the past}} = \frac{100}{1000} = .1 \qquad x = 2 \qquad P(2; .1) = .0045$$

Exercise 8

300 apples were checked for worms and a total of 60 worms were found. What is the probability of an apple with 1 worm in it if this approximates a Poisson?

Example 9--Hypergeometric.

A bag has 10 red and 20 blue marbles. If you pull out 5 without replacement, find the probability of getting 3 red and 2 blue marbles.

Solution

This is a hypergeometric since there are repeated dependent trials.

$$P(A) = \frac{{}_aC_x \, {}_bC_{n-x}}{{}_{(a+b)}C_n}$$

n = total number of items to be picked = 5
a = number of the first item in the total = 10 (reds)
b = number of the second item in the total = 20 (blues)
x = number of the first item desired = 3
$n - x$ = number of the second item desired = 2

$$P(A) = \frac{{}_{10}C_3 \, {}_{20}C_2}{{}_{30}C_5} = \frac{\frac{10!}{7!3!} \cdot \frac{20!}{18!2!}}{\frac{30!}{25!5!}} = \frac{120(190)}{142,506} = .160$$

Exercise 9

Find the probability of getting 1 red and 4 blues when 5 marbles are drawn from a bag with 8 red and 10 blue marbles.

QUESTIONS:

Practice Test

1. Find the mean and standard deviation for the following probability distribution.

x	$P(x)$
3	.26
6	.34
9	.28
12	.12

2. If 23% of all doctors are internists, find the probability that in a group of 15 doctors, 4 are internists.

3. Traffic lights with a turning arrow are red 70% of the time. If you approach 6 lights in a row, find the probability of stopping at least 3 times.

4. A company gave a drug test to its 300 employees. If the test is effective 98% of the time, find the mean and standard deviation for the number of correct drug tests.

5. 10% of the cameras sold in America are made in Europe, 70% are made in America, and 20% are made in Japan. If 6 cameras are picked at random, what is the probability that 1 was made in Europe, 3 in America, and 2 in Japan?

6. If 4% of the population catch a certain disease, find the probability that in a group of 200 people, 3 will get the disease. This approximates a Poisson.

7. A math lab has an average of 5 students per hour. Find the probability there will be 8 students in the lab for a certain hour if this approximates a Poisson.

8. An instructor graded 200 papers and found 80 errors. If a paper is picked at random, find the probability that it will have 4 errors. This approximates a Poisson.

9. In a class of 20 students, 16 made C or better and 4 made below C on the first test. If 6 students are picked to do a special project, find the probability of getting 4 with C or better and 2 below C.

CHAPTER 7
THE NORMAL DISTRIBUTION
Understanding the Normal Distribution

Introduction

A set of continuous variables where the mean, the median, and the mode are all equal is called a **normal distribution**. The graph of a normal distribution is symmetrical and approximates the bell shape shown below. The area under the bell graph is equal to 1 (or 100%). Since the mean is the same as the mode, the highest point on the bell is the mean. Since the median is the same as the mean and the median, 50% of the values are below the mean and 50% are above the mean.

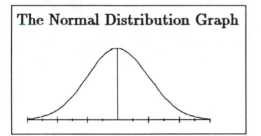

The Normal Distribution Graph

The normal distribution is used to find probabilities by finding the area under the curve. Table E gives the area under the graph from the mean to any given z score. The area under the curve is the same as the probability that a value will be between the mean and the given number. To find the probability for a normal distribution, first find the z score for the given number, x.

$$z = \frac{x - \mu}{\sigma}$$

Then look the z score up in Table E (in the back of the text or in the back of this manual) to get the area under from the mean to the given number (x). If you are looking for a probability that is between the mean and the given number, then the table value is that probability. If you are looking for another probability, the pictures given in Chart 1 tell you what to do with the table value to find the probability you want.

Steps to find the Probability for a Normal Distribution:

1. Find the z score. $z = \frac{x - \mu}{\sigma}$

2. Find the table value for the z score.
3. Draw a graph of the probability you want to find.
4. Find the graph on Chart 1 that matches your graph and find the probability you want.

Chart 1 is on the next page.

Chart 1

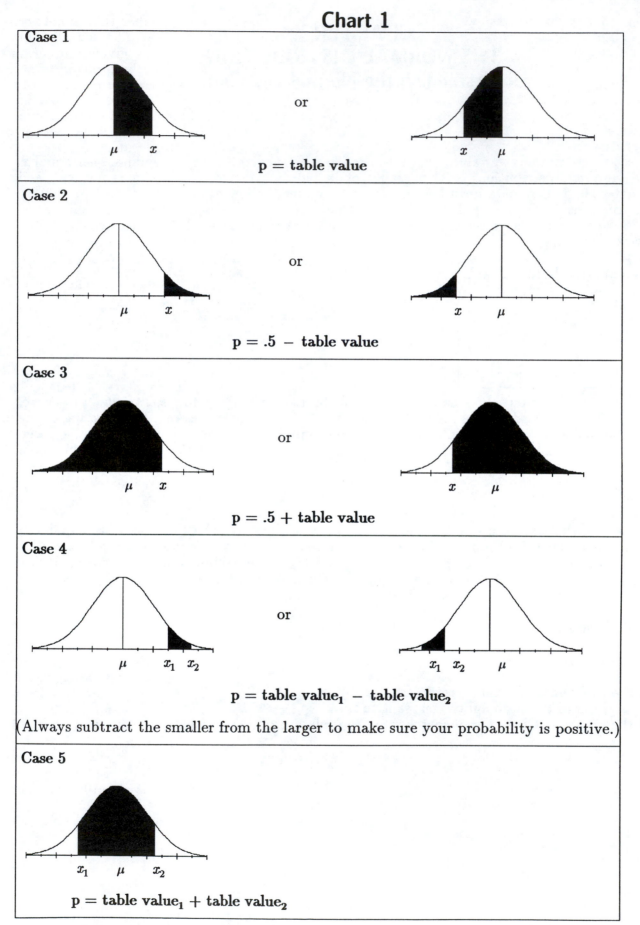

Case 1

p = table value

Case 2

p = .5 − table value

Case 3

p = .5 + table value

Case 4

p = table value₁ − table value₂

(Always subtract the smaller from the larger to make sure your probability is positive.)

Case 5

p = table value₁ + table value₂

Case 1 will be any problem that is looking for the probability of a number that is between the mean and the given number (x). Notice that normal probabilities do not find the probability of getting one specific value, but are always between two values, larger than a given number, or smaller than a given number. This is called a **continuous variable**. Case 2 is the probability of a number being larger than the given number if the given number is larger than the mean, or of a number being smaller than a number if the given number is smaller than the mean. Case 3 is the probability of a number being less than a given number that is larger than the mean or more than a given number that is smaller than the mean. Case 4 is the probability of being between two given numbers if the numbers are both larger or both smaller than the mean. Case 5 is the probability of being between 2 numbers if one is smaller and one is larger than the mean.

The Central Limit Theorem

The normal distribution can also be applied to sample means drawn from a larger population. **The Central Limit Theorem** states that the shape of the distribution of sample means taken from a population with a mean μ and standard deviation σ will approach a normal distribution and the distribution will have a mean μ and standard deviation of $\frac{\sigma}{\sqrt{n}}$. The Central Limit Theorem is used to find the probabilities concerning a sample mean instead of a single value. The steps are the same as the steps for finding probabilities for a normal distribution except the formula for z is changed.

For probabilities about a sample mean:

$$z = \frac{\bar{x} - \mu}{\frac{\sigma}{\sqrt{n}}}$$ $\bar{x}$ is the sample mean and μ is the population mean

---

If the sample is drawn without replacement from a population of small, finite size, a **correction factor** is used. The **correction factor** should be used if the sample is greater than 5% of the population.

Using the Correction Factor (if n is greater than 5% of N)

$$z = \frac{\bar{x} - \mu}{\frac{\sigma}{\sqrt{n}} \sqrt{\frac{N-n}{N-1}}}$$ n = sample size N = population size

---

Normal Approximation to the Binomial

A normal distribution can be used to approximate a binomial. A binomial probability is a repeated number of independent trials that can only have 2 outcomes. If n is large and p is not close to 0 or 1, then the normal is a fairly good approximation to the binomial probability. In a binomial probability, n is the number of trials, p is the probability of a success, and q is $1 - p$. **The normal can be used to approximate the binomial if both np and nq are larger than or equal to 5.**

Since binomials are a discrete distribution and normals are a continuous distribution, a **correction factor** is used when the normal is used to approximate the binomial. For any given value, x, .5 is added or subtracted to get the new boundaries. For instance, if $x = 10$ is used in the binomial, then $9.5 \le x \le 10.5$ is used in the normal. $x \le 10$ would be $x \le 10.5$. $x \ge 10$ would be $x \ge 9.5$ in the normal. You would either add or subtract .5 whichever would give you more area under the normal curve.

To Use the Normal to Approximate the Binomial:

1. Check to see if the normal approximation can be used. $np \geq 5$ and $nq \geq 5$.
2. Find $\mu = np$ and $\sigma = \sqrt{npq}$.
3. Use the correction factor to write the probability problem.

4. Find $z = \frac{x - \mu}{\sigma}$.

5. Find the table value for z.
6. Draw a graph of the probability you are looking for.
7. Use the graphs in Chart 1 to find the probability.

NOTES:

Checking Your Understanding

Complete this section before you do the exercises to make sure you understand the concepts. Write in the book. The answers are in the back of the book. Make a note of any questions that you wish to discuss with the instructor.

I. List the conditions that must be met in order to have a normal distribution.

II. When can the normal distribution be used to approximate the binomial probability?

III. Select the correct answer and write the appropriate letter in the space provided.

____ 1. A normal distribution has
   a. discrete variables.
   b. continuous variables.

____ 2. A binomial distribution has
   a. discrete variables.
   b. continuous variables.

____ 3. When looking for a probability concerning a sample mean, use the
   a. normal approximation to the binomial.
   b. normal distribution.
   c. Central Limit Theorem.
   d. correction factor.

____ 4. When using the normal approximation to the mean, use the
   a. Central Limit Theorem.
   b. correction factor.

QUESTIONS:

Applying Your Understanding

STUDENT: In the preceding sections, you have learned the concepts of normal probability and checked your understanding of the concepts. Study each example carefully and then try to work the following exercises. If you have any problems, see your instructor. The answers are in the back of the book.

Example 1--Normal probability.

At a certain store, the average amount spent by each customer who makes a purchase is $25 with a standard deviation of $8. If a customer is picked at random, find the probability that the amount spent will be:
(a) between $25 and $40
(b) more than $40
(c) less than $40.

Solution

(a) 1. $z = \frac{x - \mu}{\sigma} = \frac{40 - 25}{8} = 1.875 = 1.88$ Need to round to 2 places after the decimal since that is as far as the table goes.

2. table value = .4699   Go to Table E. Look down the far left column, labeled z, until you find the number before the decimal and the first digit after the decimal, 1.8. Then go across that row until you find the second digit after decimal, .08, in the top row and read the table value from that column.

3. Draw a bell shaped graph. Label the highest point of the bell the mean, 25. Since 40 is more than the mean, put a mark for 40 to the right of the mean. Since the desired probability is between 25 and 40, shade in the region between 25 and 40.

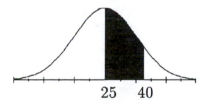

25   40

4. This picture matches Case 1, so p = table value.   p = .4699.

(b) 1. $z = \frac{x - \mu}{\sigma} = \frac{40 - 25}{8} = 1.875 = 1.88$

2. table value = .4699
3. Draw a bell shaped graph. Label the highest point of the bell 25 and 40 to the right of 25. Since the desired probability is more than 40, shade to the right of 40.

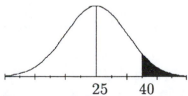

25   40

4. This picture matches Case 2, so p = .5 − table value = .5 − .4699 = .0301.

(c) 1. $z = \frac{x - \mu}{\sigma} = \frac{40 - 25}{8} = 1.875 = 1.88$

2. table value = .4699
3. Draw a bell shaped graph. Label the highest point of the bell 25 and 40 to the

right of 25. Since the desired probability is less than 40, shade to the left of 40.

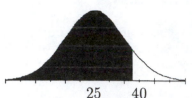

25   40

4. This picture matches Case 3, so p = .5 + table value = .5 + .4699 = .9699.

Exercise 1

A machine is set to fill bags with 5 pounds of sugar, but the mean is found to be 5.21 pounds with a standard deviation of .16. Find the probability of a bag of sugar having:
(a) between 5.21 and 5.41 pounds of sugar.
(b) more than 5.41 pounds of sugar.
(c) less than 5.41 pounds of sugar.

Example 2--Normal probability.

The mean time to complete a certain psychology exam is 34 minutes with a standard deviation of 8. What is the probability of a student taking:
(a) more than 28 minutes
(b) between 15 and 28 minutes
(c) between 28 and 41 minutes?

Solution

(a) 1. $x = \dfrac{28 - 34}{8} = -0.75$ (NOTE: make sure you are always subtracting the mean from the given x.)

2. table value = .2734   Find 0.7 under the column labeled z and then go across the top row until you find .05. The table value is a probability so it is always positive.

3. Draw a bell shaped graph. Label the mean, 34, at the highest point. Put 28 to the left of 34. Shade everything to the right of 28.

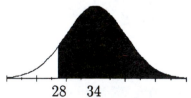

28   34

4. This matches Case 3, p = .5 + table value = .5 + .2734 = .7734.

(b) 1. Since there are two given amounts, find two different z's and two table values.

$z_1 = \dfrac{15 - 34}{8} = -2.375 = -2.38$ $z_2 = \dfrac{28 - 34}{8} = -0.75$

2. table value$_1$ = .4913   table value$_2$ = .2734

3. Draw a bell shaped graph. Label the highest point 34, 28 to the left of 34, and 15 to the left of 28.

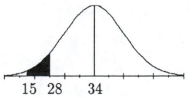

15 28   34

4. This matches Case 4, p = table value$_1$ – table value$_2$. (Take the smaller value from the larger to make sure you answer is positive.)
   p = .4913 – .2734 = .2179

(c)   1. $z_1 = \dfrac{28 - 34}{8} = -0.75$ $z_2 = \dfrac{41 - 34}{8} = 0.88$

2. Table value$_1$ = .2734   table value$_2$ = .3106
3. Shade the area between 28 and 41.

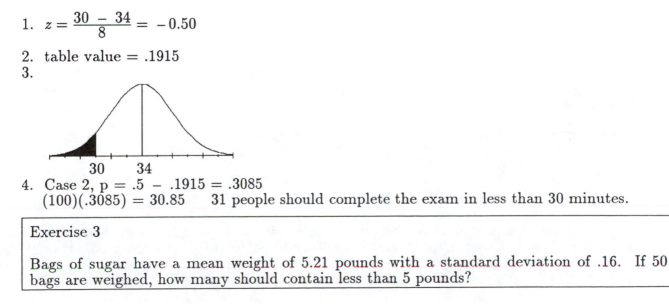

28   34   41

4. This matches Case 5, p = table value$_1$ + table value$_2$ = .2734 + .3106 = .5840.

Exercise 2

If the mean weight of bags of sugar is 5.21 with a standard deviation of .16, what is the probability of a bag of sugar with
(a) less than 4.91 pounds of sugar?
(b) more than 5.03 pounds of sugar?
(c) between 5.03 and 4.91 pounds of sugar?
(d) between 4.91 and 5.69 pounds of sugar?

Example 3--Normal probability.

The mean time to complete a certain psychology exam is 34 minutes with a standard deviation of 8. If 100 students take the exam, how many should finish in less than 30 minutes?

Solution

Find the probability of completing the exam in less than 30 minutes then multiply the probability times the number of people taking the exam to find out how many will complete in less than 30 minutes.

1. $z = \dfrac{30 - 34}{8} = -0.50$

2. table value = .1915
3.

30   34

4. Case 2, p = .5 – .1915 = .3085
    (100)(.3085) = 30.85   31 people should complete the exam in less than 30 minutes.

Exercise 3

Bags of sugar have a mean weight of 5.21 pounds with a standard deviation of .16. If 50 bags are weighed, how many should contain less than 5 pounds?

Example 4--Finding x for a given probability.

A pizza restaurant delivers in a mean time of 18.2 minutes with a standard deviation of 3 minutes. They want to advertise that they will not charge for a pizza that takes longer than a certain length of time to deliver. If they want to give away no more than 5% of the time, how many minutes should they claim they can deliver in?

Solution

Since this problem gives the probability and asks for the x, it is worked backward from the probability problem. 5% can be thought of as the probability of .05.

1. Draw a picture to determine what was done to the table value to get the probability .05. The problem is looking for a number that is larger than the mean, 18.2, because they are only going to give away pizzas if their time is too large, so label 18.2 at the highest point and label x to the right of 18.2. Shade in to the right of x since you are looking the greater than x.

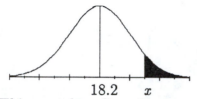

2. This matches Case 2, so p = .5 − table value. The probability is .05, so solve for the table value.

    p = .5 − table value
    .05 = .5 − table value   Subtract .5 from both sides of the equation.
    −.45 = − table value   Divide both sides by −1.
    .45 = table value

3. Find the z value by looking at Table E. This time look in the middle of the table values until you find the value that is closest to .45. Then look to the left column to find the z value. z = 1.64 or 1.65 since .45 is between .4495 and .4505. Look at the graph. If x is to the left of the mean, the z will be negative. If x is to the right, z is positive. z = +1.65

4. $z = \frac{x - \mu}{\sigma}$

    $1.65 = \frac{x - 18.2}{3}$ To solve for x, multiply both sides of the equation by 3.

    $4.95 = x - 18.2$ Add 18.2 to both sides of the equation.
    $23.15 = x$

Pizzas will be delivered in over 23 minutes 5% of the time.

Exercise 4

A teacher gives a test and the results are normally distributed with a mean of 36 and a standard deviation of 5. If the teacher wants to give the top 10% of the class A's, what should the cutoff be for an A?

Example 5--Finding x for a given probability.

A company gives an employment test to all applicants for a job. The results of the test are normally distributed with a mean score of 124 and a standard deviation of 16. If only the top 75% of the applicants are to be interviewed, what score must an applicant have to be interviewed?

Solution

1. Draw a picture. Label the mean, 124. Since you are looking for the top 75% of the applicants, the x will be smaller than the mean. Shade to the right of the x.

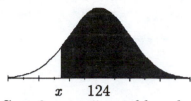

x   124

2. Case 3, p = .5 + table value   p = .75
   .75 = .5 + table value   Subtract .5 from both sides of the equation.
   .25 = table value

3. z for the table value closest to .25 is 0.67. Since x is to the left of the mean, $z = -0.67$.

4. $z = \dfrac{x - \mu}{\sigma}$

   $-0.67 = \dfrac{x - 124}{16}$ Multiply both sides of the equation by 16.

   $-10.72 = x - 124$ Add 124 to both sides of the equation.
   $113.28 = x$ All scores over 113 should be interviewed.

Exercise 5

A teacher gives a test and the results are normally distributed with a mean of 36 and a standard deviation of 5. If the teacher wants to give the lowest 20% F's, find the cutoff score for an F.

Example 6--Central Limit Theorem

The mean time to complete a certain psychology exam is 34 minutes with a standard deviation of 8. If a class of 30 take the exam, find the probability that the sample mean will be more than 30 minutes?

Solution

Since you are looking for a probability involving a sample mean, you use the Central Limit Theorem. Work like the other normal probabilities except

$z = \dfrac{\bar{x} - \mu}{\frac{\sigma}{\sqrt{n}}}$.

1. $z = \dfrac{30 - 34}{\frac{8}{\sqrt{30}}} = -2.74$

2. table value = .4969
3. Draw the graph.

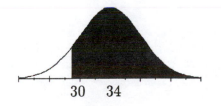

30   34

4. Case 3, p = .5 + table value = .5 + .4969 = .9969

Exercise 6

The mean weight for bags of sugar is 5.21 pounds with a standard deviation of .16. If a sample of 40 bags is taken, what is the probability that the mean weight will be less than 5.15 pounds?

Example 7--Normal Approximation to the Binomial.

A doctor's office claims it keeps appointments on time 65% of the time. If 120 patients are seen in one week, find the probability that at least 65 patients will be seen on time.

Solution

This a binomial since there are only two choices and there are repeated independent trials.
$n = 120$
$p = .65$
$q = 1 - .65 = .35$
1. See if the normal approximation can be used. $np = 120(.65) = 78$ $nq = 120(.35) = 42$
    np and nq are both greater than 5, so the normal approximation can be used.
2. $\mu = np = 78$

$$\sigma = \sqrt{npq} = \sqrt{120(.65)(.35)} = \sqrt{27.3} = 5.22$$
3. The desired probability is at least 65. $x \geq 65$. Using the correction factor, $x \geq 64.5$.

4. $z = \dfrac{x - \mu}{\sigma} = \dfrac{64.5 - 78}{5.22} = -2.59$

5. table value = .4952
6. Draw a graph. Shade to the right of 64.5.

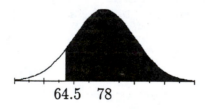

64.5   78

7. Case 3, p = .5 + table value = .5 + .4952 = .9952

Exercise 7

Using the data in Example 6, find the probability that less than 85 patients will be seen on time.

QUESTIONS:

Practice Test

1. A light bulb is advertised as lasting an average of 1000 hours with a standard deviation of 150 hours. Find the probability of buying a light bulb that will last:
   a. between 1000 and 1250 hours

   b. less than 1250 hours

   c. more than 1250 hours

   d. between 901 and 1000 hours

   e. less than 901 hours

   f. more than 901 hours

   g. between 901 and 1250 hours

   h. between 649 and 901 hours

2. A college gives an entrance exam which results in a mean score of 247 with a standard deviation of 28. If only the upper 70% of the applicants are to be admitted, what should the cutoff score be?

3. A certain diet claims to lower cholesterol by an average of 60 points with a standard deviation of 7. Find the probability that the lowered amount will be more than 75 points.

4. Mensa is an organization that accepts members who score in the top 2% of the population on standard intelligence tests. If the mean IQ is 100 with a standard deviation of 15, what should the minimum IQ score be to be accepted by Mensa?

5. The mean IQ for the general public is 100 with a standard deviation of 15. If 20 people are tested, what is the probability the average IQ will be above 108?

6. A librarian knows that 14% of the borrowers will not return the books by their due date. If a sample of 75 borrowers is tested, find the probability that more than 15 will not return the books on time.

CHAPTER 8
CONFIDENCE INTERVALS AND SAMPLE SIZE
Understanding Confidence Intervals and Sample Size

Means

When it is impossible or impractical to work with a whole population, there are two estimators that can be used for the population mean. The **point estimate** for the population mean is the sample mean. The **interval estimate** states two values that the population mean should fall between. The intervals are called **confidence intervals**. An alpha level is picked to correspond to what confidence is desired. For example, a 99% confidence interval would be $\alpha = 1 - .99 = .01$. If a 95% confidence interval is $20 \leq \mu \leq 28$, then you can be 95% confident that the population mean is between 20 and 28. A z value is used to find the confidence interval if the population standard deviation is known or if the sample size is greater than or equal to 31. A t value is used if the population standard deviation is not known and the sample size is less than 31. The z and the t values are found on tables found in the back of the text or this manual.

The confidence interval for the population mean, if the population standard deviation is known or if $n \geq 31$:

$$\bar{x} - z_{\alpha/2}\left(\frac{\sigma}{\sqrt{n}}\right) \leq \mu \leq \bar{x} + z_{\alpha/2}\left(\frac{\sigma}{\sqrt{n}}\right) \qquad z_{\alpha/2} = \text{two tailed } z \text{ from Table F for the desired confidence interval}$$

The confidence interval for the population mean, if the sample standard deviation is known or if $n < 31$:

$$\bar{x} - t_{\alpha/2}\left(\frac{s}{\sqrt{n}}\right) \leq \mu \leq \bar{x} + t_{\alpha/2}\left(\frac{s}{\sqrt{n}}\right) \qquad t_{\alpha/2} = \text{two tailed } t \text{ from Table F for the desired confidence interval and df} = n - 1$$

To decide the minimum sample size to make an accurate estimate, the population standard deviation must be known. The level of confidence and the maximum error of estimate must be decided upon. The **maximum error estimate**, E, is the largest amount of error in the estimated mean the person doing the study wishes to have.

Minimum sample size needed for an interval estimate for the population mean is

$$n = \left(\frac{z_{\alpha/2}\,\sigma}{E}\right)^2 \qquad \textbf{Round up}$$

E is the maximum error of estimate
$z_{\alpha/2}$ = two tailed z from Table F for the desired confidence interval

---

Proportions

The point estimate for a population proportion, p, is the sample proportion, $\hat{p}$. To find a confidence interval for a population proportion, $n\hat{p}$ and $n\hat{q}$ must be greater than or equal to 5 ($\hat{q} = 1 - \hat{p}$). When $\hat{p}$ is unknown, use $\hat{p} = .5$.

Confidence interval for the population proportion when $n\hat{p}$ and $n\hat{q} \geq 5$ $(\hat{q} = 1 - \hat{p})$

$$\hat{p} - z_{\alpha/2}\sqrt{\frac{\hat{p}\hat{q}}{n}} \leq p \leq \hat{p} + z_{\alpha/2}\sqrt{\frac{\hat{p}\hat{q}}{n}} \qquad \begin{array}{l} z_{\alpha/2} = \text{two tailed } z \text{ from Table F} \\ \qquad \text{for the desired confidence interval} \end{array}$$

Minimum sample size needed for an interval estimate for the population proportion is

$$n = \hat{p}\hat{q}\left(\frac{z_{\alpha/2}}{E}\right)^2 \qquad \textbf{Round up}$$

E = maximum error of estimate
$\hat{p}$ = sample proportion
$\hat{q} = 1 - \hat{p}$
z = two tailed value for the desired confidence interval from Table F

NOTES:

Checking Your Understanding

Complete this section before you do the exercises to make sure you understand the concepts. Write in the book. The answers are in the back of the book. Make a note of any questions you wish to discuss with the instructor.

Fill in the blank with the best answer:

1. The best point estimate for the population mean is the _____.

2. The best point estimate for the population proportion is the _____.

3. If the confidence interval desired is 95%, alpha would be _____.

4. The z or t value found for a confidence interval is a(n) _____ tailed value.

5. The largest error in the estimate is the _____.

QUESTIONS:

QUESTIONS:

Applying Your Understanding

STUDENT: In the preceding sections, you learned the concepts of confidence intervals and sample size and checked your understanding of the concepts. In this section, you will apply your understanding of the concepts. Study each example carefully and then try to work the following exercise. If you have any problems, see your instructor. The answers to the exercises are in the back of the book.

Example 1--Confidence interval for a population mean.

If a sample of 48 has a sample mean of 416 and a standard deviation of 15, find a 95% confidence interval for the population mean.

Solution

Use the z since $n = 48 \geq 31$. $\alpha = 1 - .95 = .05$, two tailed $z = 1.960$

$$\bar{x} - z_{\alpha/2}\left(\frac{\sigma}{\sqrt{n}}\right) \leq \mu \leq \bar{x} + z_{\alpha/2}\left(\frac{\sigma}{\sqrt{n}}\right)$$

$$416 - 1.96\left(\frac{15}{\sqrt{48}}\right) \leq \mu \leq 416 + 1.96\left(\frac{15}{\sqrt{48}}\right)$$

$416 - 4.2 \leq \mu \leq 416 + 4.2$
$411.8 \leq \mu \leq 420.2$

There is a 95% probability that the population mean is between 411.8 and 420.2.

Exercise 1

If a sample of 45 has a sample mean of 83 and a standard deviation of 3.5, find the 95% confidence interval for the population mean.

Example 2--Confidence interval for a population mean.

Find a 99% confidence interval for the population mean if a sample of 28 has a mean of 212 and a standard deviation of 21.

Solution

Use a t since $n = 28 < 31$ and the sample standard deviation is given. $\alpha = 1 - .99 = .01$, df $= 28 - 1 = 27$, $t_{\alpha/2} = 2.771$

$$\bar{x} - t_{\alpha/2}\left(\frac{s}{\sqrt{n}}\right) \leq \mu \leq \bar{x} + t_{\alpha/2}\left(\frac{s}{\sqrt{n}}\right)$$

$$212 - 2.771\left(\frac{21}{\sqrt{28}}\right) \leq \mu \leq 212 + 2.771\left(\frac{21}{\sqrt{28}}\right)$$

$212 - 11 \leq \mu \leq 212 + 11$
$201 \leq \mu \leq 223$

Exercise 2

Find a 90% confidence interval for the population mean if the sample mean is 1462 with a standard deviation 42 for a sample of 20.

Example 3--Sample size for a mean.

Find the minimum sample size needed to estimate a population mean if the population standard deviation is 24, the confidence interval desired is 95%, and a maximum error of estimate is 5.

Solution

$\alpha = 1 - .95 = .05$, two tailed $z = 1.645$ $n = \left(\frac{z\sigma}{E}\right)^2 = \left[\frac{(1.645)(24)}{5}\right]^2 = 62.3$ Round up to 63

Exercise 3

Find the minimum sample size to estimate a population mean if the population standard deviation is 5, the confidence interval is 95%, and the maximum error of estimate is 2.

Example 4--Confidence interval for a population proportion.

Find a 99% confidence interval for the population proportion if a sample of 200 had a sample proportion of 42%.

Solution

$n\hat{p} = 200(.42) = 84$ $n\hat{q} = 200(.58) = 116$ both are greater than 5
$\alpha = 1 - .99 = .01$ two tailed $z = 2.576$

$$\hat{p} - z_{\alpha/2}\sqrt{\frac{\hat{p}\hat{q}}{n}} \leq p \leq \hat{p} + z_{\alpha/2}\sqrt{\frac{\hat{p}\hat{q}}{n}}$$

$$.42 - 2.576\sqrt{\frac{.42(.58)}{200}} \leq p \leq .42 + 2.576\sqrt{\frac{.42(.58)}{200}}$$

$.42 - .09 \leq p \leq .42 + .09$
$.33 \leq p \leq .51$

You can be 99% confident that the population proportion is between .33 and .51.

Exercise 4

Find a 90% confidence interval for the population proportion if a sample of 98 had a sample proportion of 21%.

Example 5--Sample size for a proportion.

Find the minimum sample size required to find the 90% interval estimate for a population proportion if you want to be accurate within 2%.

Solution

Since $\hat{p}$ is not given, let $\hat{p} = .5$ and $\hat{q} = 1 - \hat{p} = .5$ $\alpha = 1 - .90 = .10$, two tailed $z = 1.645$

$$n = \hat{p}\hat{q}\left(\frac{z_{\alpha/2}}{E}\right)^2 = (.5)(.5)\left(\frac{1.645}{.02}\right)^2 = 1691.2656$$ Round up to 1692.

Exercise 5

Find the minimum sample size to find the 90% confidence interval for a population proportion if you want to be accurate to 4%.

Practice Test

1. Find the 90% confidence interval for a population mean if the sample mean is 21.6 with a standard deviation of 2.8 for a sample of 35.

2. Find the 95% confidence interval for a population mean if the sample mean is 146 with a standard deviation of 12 for a sample of 18.

3. Find the minimum sample size to estimate a population mean if the population standard deviation is 5.6, the confidence interval is 99%, and the maximum error of estimate is 4.

4. Find a 90% confidence interval for the population proportion if a sample of 106 had a sample proportion of 21%.

5. Find the minimum sample size to find a 95% interval estimate for a population proportion if a previous sample had a sample proportion of 35% and you want to be accurate within 3% of the true proportion.

CHAPTER 9
HYPOTHESIS TESTING
Understanding Hypothesis Testing

Introduction

Hypothesis testing is used to compare a sample to the population or to compare two samples. The two hypotheses that are tested are called the null hypothesis and the alternate hypothesis. The **null hypothesis (H_0)** states that there is no difference between 2 groups being compared. The **alternate hypothesis (H_1)** states that there is a specific difference between the two groups. The hypothesis test is used to decide whether to accept the null or the alternate hypothesis. The null and alternate hypotheses can have three different forms.

H_0: $=$ H_0: $\leq$ H_0: $\geq$
H_1: $\neq$ H_1: $>$ H_1: $<$

Notice that the null hypothesis always has the equal involved ($=$, $\leq$, or $\geq$). The first case (H_0: $=$ and H_1: $\neq$) is called a **two tailed** test because the second parameter could either be larger or smaller than the first. Therefore, there are two ways to satisfy the alternate. The second and third cases are called **one tailed** tests because there is only one way to satisfy the alternate.

A table of some common phrases and the symbol to use is given below.

$>$	$<$
is greater than	is less than
is more than	is below
is larger than	is lower than
is longer than	is shorter than
is bigger than	is smaller than
is better than	is reduced

$\geq$	$\leq$
is greater than or equal	is less than or equal to
is at least	is not more than
is not less than	is at most

$=$	$\neq$
is equal to	is not equal to
is exactly	is different
has not changed	has changed
is the same as	is not the same as

Statistical Tests

A **statistical test** uses data from a sample to decide whether the null hypothesis is to be accepted or rejected. A **test value** is the number obtained from the statistical test. When deciding to accept or reject the null, there are two errors that can be made. A **Type I error** would be made if a null hypothesis is rejected when it is true. A **Type II error** would be made if the null is not rejected when is is false. The **level of significance** is the maximum probability of committing a Type I error. This probability is called **alpha (α)**. The two most common levels of significance are .05 or .01. At the .01 level, the probability of making a Type I error is 1% or there is a 99% chance that a Type I error will not be made. The **critical** or **rejection region** is the area that indicates there is a significant difference so the null hypothesis should be rejected. The **critical value** separates the critical region from the non-critical region. The non-critical region is where the null is accepted.

Steps to Hypothesis Testing
1. State the hypotheses.
2. Find the critical value or values from the table. Draw a graph and label the critical values.
3. Compute the test value.
4. Make a decision to reject or not to reject the null.
5. Summarize the results.

Z Test

The *z* test is used when the population is normally distributed and either the population standard deviation is known or the sample size is greater than or equal to 30 ($n \geq 30$). Since the *z* test is used when the population is normally distributed, Table E (in the back of the text or this manual) is used to find the critical values for specific α's.

The test value for the *z* test:

$$z = \frac{\bar{x} - \mu}{\frac{\sigma}{\sqrt{n}}} \qquad \text{or} \qquad z = \frac{\bar{x} - \mu}{\sigma}\sqrt{n} \quad \text{is easier to compute}$$

$\bar{x}$ = sample mean
μ = population mean (always the value in the hypotheses)
σ = standard deviation
n = sample size

To find the critical values for the *z* test:
1. Draw a bell shaped curve and indicate the appropriate area.
    a. If the alternate hypothesis is =, the critical region will be two tails, one to the left of the mean and one to the right. α must be divided by 2 for a two tailed test.
    b. If the alternate hypothesis is <, critical region will be the tail to the left.
    c. If the alternate hypothesis is >, the critical region will be the tail to the right.

2. Find $.5 - \alpha$ for either one tailed test or $.5 - \frac{\alpha}{2}$ for the two tailed test.

3. Look at the table values in Table E to find the table value closest to the number found in step 2. Find the *z* value for that table value. The *z* value will be the critical value. For the left tails the critical value is negative, for right tail the critical value is positive.

Since .05 and .01 are the most commonly used levels of significance, the following chart is useful in finding their critical values.

Chart 2
Critical Values for the *z* Test

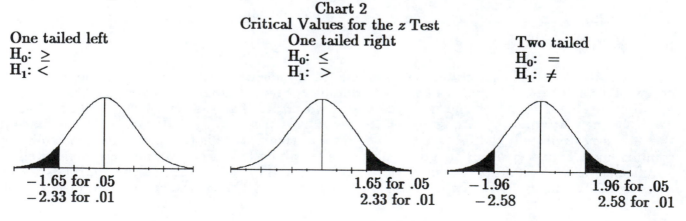

One tailed left
$H_0: \geq$
$H_1: <$

One tailed right
$H_0: \leq$
$H_1: >$

Two tailed
$H_0: =$
$H_1: \neq$

-1.65 for .05
-2.33 for .01

1.65 for .05
2.33 for .01

-1.96
-2.58

1.96 for .05
2.58 for .01

Table F can be used to find the z values. Find the column for the desired level of significance for either one or two tailed test. Read the bottom row of numbers [labeled (z) ∞] for the desired column. Table F is the t table but when $n \geq 30$, the t values and the z values are approximately equal.

T test

The **t test** is used when the population is normally distributed, but the population standard deviation is not known and the $n < 30$. In the t test the sample standard deviation is used, and the critical values are found on Table F for df $= n - 1$. Df are called the **degrees of freedom**.

The test value for the t test:

$$t = \frac{\bar{x} - \mu}{\frac{s}{\sqrt{n}}} \qquad \text{or} \qquad z = \frac{\bar{x} - \mu}{s}\sqrt{n} \quad \text{is easier to compute}$$

$\bar{x}$ = sample mean
μ = population mean (**always the value in the hypotheses**)
s = standard deviation
n = sample size

To find the critical values for the t test on Table F, find the desired level of significance for one or two tailed test. Look at the left hand side to find df $= n - 1$ and read that row under the correct column. The table value is the critical value. Note that on Table F, all df's larger than 29 are grouped together. If df > 29, then the t value and the z are very close together so the bottom row is used for the t test if df > 29 and for the z test.

Test for Proportions

If a sample is to be compared to a normal population proportion, a z test can be used if $np \geq 5$ and $nq \geq 5$.

The test value for the z test for proportion:

$$z = \frac{x - np}{\sqrt{npq}}$$

x = number with the desired trait in the sample
n = sample size
p = proportion in the sample
$q = 1 - p$

P-values

The **P-value** for a hypothesis test is the same as the probability of the sample mean occurring if the null hypothesis is true. The P-value is found the same as finding the normal probability. If the P-value is greater than α, accept the null. If the P-value is less then α, reject the null. For two tailed tests, the P-value is two times the probability.

Note that all the tests in this chapter were applied to normal populations. Tests for populations that are not normally distributed are called Non-Parametric and will be discussed in later chapters.

Summary for Comparing a Sample to a Normal Population

Steps to Hypothesis Testing
1. State the hypotheses.
2. Find the critical value or values from the table. Draw a graph and label the critical values.
3. Compute the test value.
4. Make a decision to reject or not to reject the null.
5. Summarize the results.

COMPARING A MEAN

I. If the population standard deviation
   is known or if $n \geq 30$.
   z test
   Test statistic:

$$z = \frac{\bar{x} - \mu}{\frac{\sigma}{\sqrt{n}}} \quad \text{or} \quad z = \frac{\bar{x} - \mu}{\sigma} \sqrt{n}$$

$\bar{x}$ = sample mean
μ = population mean
σ = standard deviation
n = sample size
Critical value: Table F, bottom row $[(z) \infty]$

II. If the sample standard deviation is
   known and if $n < 30$.
   t test
   Test statistic:

$$t = \frac{\bar{x} - \mu}{\frac{s}{\sqrt{n}}} \quad \text{or} \quad t = \frac{\bar{x} - \mu}{s} \sqrt{n}$$

$\bar{x}$ = sample mean
μ = population mean
s = standard deviation
n = sample size
Critical value: Table F, df = $n - 1$

COMPARING A PROPORTION

np and nq must both be ≥ 5

z test
Test statistic:

$$z = \frac{x - np}{\sqrt{npq}}$$

x = number with the desired trait in sample
n = sample size
p = proportion in the sample
$q = 1 - p$
Critical value: Table F, bottom row $[(z) \infty]$

Checking Your Understanding

Complete this section before you do the exercises to make sure you understand the concepts. Write in the book. The answers are in the back of the book. Make a note of any questions that you wish to discuss with the instructor.

I. If you want to test to see if a sample is greater than the population mean of 32, write the null and alternate hypotheses.

II. If you want to test a statement that a sample is at most 17, write the null and alternate hypotheses.

III. If you want to test that a procedure improved completion time on a sample where the population mean completion was 12.3 minutes, write the null and alternate hypotheses.

IV. To test that a cheaper gasoline did not change your gas mileage when your gas mileage before was 18, write the hypotheses.

V. Select the correct answer and write the appropriate letter in the space provided.

_____ 1. If the population standard deviation is not known and n is less than 30, use the
   a. t test.
   b. z test.
   c. test for proportions.
   d. P-value.

_____ 2. $\mu \geq 16$ would be the
   a. null hypothesis.
   b. alternate hypothesis.

_____ 3. $\mu <$ would be the
   a. null hypothesis.
   b. alternate hypothesis.

_____ 4. The test for proportions uses the z test if
   a. $np \geq 5$.
   b. $nq \geq 5$.
   c. np and $nq \geq 5$.
   d. $np \leq 5$.

_____ 5. The number that divides the critical region for the non-critical region is the
   a. P-value.
   b. critical value.
   c. level of significance.
   d. alpha.

_____ 6. To accept a null hypotheses that is false is a
   a. Type I error.
   b. Type II error.

_____ 7. To reject a null hypothesis which is true is a
   a. Type I error.
   b. Type II error.

QUESTIONS:

Applying Your Understanding

STUDENT: In the preceding sections, you learned the concepts of hypothesis testing and checked your understanding of the concepts. In this section, you will apply your understanding of the concepts. Study each example carefully and then try to work the following exercise. If you have any problems, see your instructor. The answers are in the back of the book.

Example 1--z test.

The mean grade point average for one college is 2.45 with a standard deviation of .69. An engineering professor believes that engineering majors have a higher grade point average than the college's mean. A sample of 20 engineering majors had a mean grade point average of 2.65. Test the professor's claim at the .01 level of significance.

Solution

1. State the hypotheses.
   The professor's belief that engineering students would have a higher mean is the hypothesis statement $\mu > 2.45$. The opposite statement is $\mu \leq 2.45$. Since the second statement has an "equal," the second statement is the null hypothesis.
   H_0: $\mu \leq 2.45$
   H_1: $\mu > 2.45$
2. Find the critical value. Draw a graph and label the critical value.
   Decide if this is a t or a z test. Since the standard deviation is given at the same time as the population mean, the standard deviation must be the population standard deviation. If the population standard deviation is given, use the z test. The level of significance is .01 and this is a one tailed test. ($\leq$ and $\geq$ are one tailed tests, $=$ is the only two tailed test). Look at Table F. Find the column at the top that is labeled .01, one tailed. Read the value at the bottom of this column for a z test. The critical value = 2.326. Draw a bell shaped graph. This is a right tail test since the alternate hypothesis is $>$. Shade in a tail on the right. **Look at the alternate hypothesis to decide where to shade on the graph ($>$ is right, $<$ is left, $\neq$ is both tails).** Label the cutoff +2.326 since it is on the right.

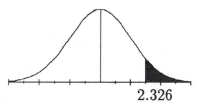

2.326

3. Compute the test statistic.

   $z = \dfrac{\overline{x} - \mu}{\sigma}\sqrt{n} = \dfrac{2.65 - 2.45}{.69}\sqrt{20} = 1.296$ (Make sure you are subtracting the value that is in the null hypothesis; this is always the population mean.)

4. **If the test statistic falls in the white region on the graph, accept the null. If the test statistic falls in the shaded region on the graph, accept the alternate.** Accept the null.
5. There is not enough evidence to accept the claim that engineering majors have a higher grade point average than the college's mean.

Exercise 1

A company manager claims that the mean number of items produced per shift is 145 with a standard deviation of 8. The supervisor believes each shift produces more than 145 items. In a sample of 25 shifts, the mean number of items produced was 150. Does each shift produce more than 145 items? Test at the .01 level.

Example 2--z test.

Last year, a grocery store had a mean of $1850 in daily sales. This month a new advertising approach was used. The store manager wants to know if the new advertising had any effect on the daily sales. If this month sales had a mean of $1780 for 30 days with a standard deviation of $150, did the new advertising affect the daily sales at the .05 level?

Solution

1. State the hypotheses.
   Since you want to know if there was any effect, test $\mu = 1850$. The opposite statement is $\mu \neq 1850$.
   $H_0: \mu = 1850$
   $H_1: \mu \neq 1850$
2. Find the critical values. Draw and label the critical values.
   This is a z since $n \geq 30$. Look on Table F for two tail and .05. Look at the bottom row.
   $z = \pm 1.960$

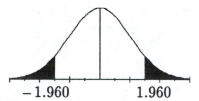

$$-1.960 \qquad 1.960$$

3. Compute the test statistic.

$$z = \frac{\bar{x} - \mu}{\sigma}\sqrt{n} = \frac{1780 - 1850}{150}\sqrt{30} = -2.556$$

4. Accept the alternate since the test statistic falls in the shaded region.
5. There was an effect on the daily sales.

Exercise 2

The mean time before a certain headache remedy starts to work is 12 minutes with a standard deviation of 3. A new coating is used to help in swallowing the pill. A sample of 18 people using the pills with the new coating showed the mean time before it started to work was 14 minutes. Is there a difference with the new coating? Test at $\alpha = .01$.

Example 3-- t test.

A consumer tested 18 bottles of a soft drink and found a sample mean of 15.8 ounces with a standard deviation of .4 ounces. If the bottles are supposed to contain 16 ounces, is the consumer being cheated?

Solution

1. State the hypotheses.
   Since the consumer is cheated if the mean is less than 16 ounces, $\mu < 16$ is one hypothesis. $\mu \geq 16$ is the opposite statement.
   $H_0: \mu \geq 16$
   $H_1: \mu < 16$
2. Find the critical value. Draw the graph and label the critical value.
   Since the standard deviation is given with the sample it is a sample standard deviation. If the sample standard deviation is given and $n < 30$, it is a t test, df $= n - 1 = 18 - 1 = 17$. The level of significance is not given so use either .01 or .05. If .05 is used, find the column that is labeled one tailed, .05 at the top and look down that column until you

are across from df = 17. $t = -1.740$ (negative since $<$ is a left tail).

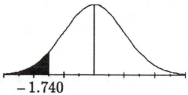

-1.740

3. Compute the test statistic. $t = \dfrac{\bar{x} - \mu}{s} \sqrt{n} = \dfrac{15.8 - 16}{.4}\sqrt{18} = -2.121$

4. Reject the null.
5. The mean is less than 16 ounces and the consumer is being cheated.

> **Exercise 3**
>
> An ambulance company claims that it responds to emergencies in less than 10 minutes. A sample of 12 calls had a mean response time of 9.6 minutes with a sample standard deviation of .9. Test at the .05 level.

Example 4-- Proportions.

A college professor feels that females are doing better in a certain math class than males. The college has 52% passing rate in that particular course. If 27 students pass the math class and 16 are females, is the proportion of females that passed higher than the proportion of the population that passed? Test at the .01 level.

Solution

To see if you can use the z test for proportions, both np and nq must be at least 5.
$n = $ sample size $= 27$ $p = $ population proportion $= .52$ $np = 27(.52) = 14.04$
$q = 1 - p = 1 - .52 = .48$ $nq = 27(.48) = 12.96$
You can use the z test since 14.04 and 12.96 $\geq$ 5.
1. State the hypotheses.
    H_0: $p \leq .52$
    H_1: $p > .52$
2. Find the critical values. Draw a graph and label the critical values.
    one tailed test, .01 level, $z = 2.326$

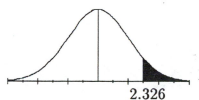

2.326

3. Compute the test statistic. $z = \dfrac{x - np}{\sqrt{npq}} = \dfrac{16 - 14.04}{\sqrt{27(.52)(.48)}} = .755$

4. Accept the null.
5. There is not enough evidence to claim that females do better in that class.

> **Exercise 4**
>
> A salesperson claims that 40% of the people that come into the store make a purchase. To test this claim, the store manager selected a sample of 100 people in the store and found that 37 made a purchase. At the .01 level, is the salesperson's claim correct?

Example 5--P-value.

Find the P-value for example 1.

Solution

Find P-value using the four steps in Chapter 8 to find the probability for a normal distribution. Find the probability greater than 2.65.

1. $z = \dfrac{\bar{x} - \mu}{\sigma}\sqrt{n} = \dfrac{2.65 - 2.45}{.69}\sqrt{20} = 1.296 = 1.30$

2. table value = .4032

3.

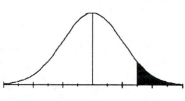

4. Case 2, p = .5 − .4032 = .0968
   Accept the null if p-value $\geq \alpha$. Accept the alternate if p-value $< \alpha$.
   At $\alpha = .01$ level, accept the null.

Exercise 5

Find the P-value for exercise 3.

Example 6--P-value.

Find the P-value for Example 2.

Solution

1. $z = \dfrac{1780 - 1850}{150}\sqrt{30} = -2.556 = -2.56$

2. table value = .4948

3.

or

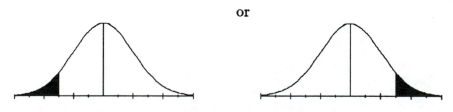

4. p = .5 − .4948 = .0052
   For a two tailed test, multiply the p times 2.   (.0052)(2) = .0104
   P-value = .0104
   At $\alpha = .05$ level, reject the null.
   (Note the null would be accepted at the .01 level.)

Exercise 6

Find the P-value for exercise 2.

Practice Test

1. An emergency team responds to calls in a mean time of 8.4 minutes with a population standard deviation of .7. A new dispatching system is being tested to see if the response time can be lowered. If a sample of 18 calls with the new system were answered in a mean time of 7.8 minutes, test at the .05 level.

2. A car gets a mean of 18 mpg. Test to see if a cheaper grade of gasoline made any difference in the gas mileage at the .01 level. A sample of 38 fillups with the cheaper gas got a mean of 17.8 mpg and a standard deviation of .6.

3. Find the p-value for problem 2.

4. A copying machine is advertised as doing at least 25,000 copies before requiring service. A consumer group wants to test this claim. If 12 customers had a mean of 24,800 copies with a standard deviation of 180, test at the .01 level.

5. A certain lottery claims that 10% of its tickets win some prize. A sample of 50 tickets had 3 winners. At the .01 level, test the claim that the proportion is 10%.

CHAPTER 10
TESTING THE DIFFERENCE BETWEEN MEANS AND PROPORTIONS
Understanding Testing the Differences

Introduction

Sometimes it is necessary to compare two sample means or sample proportions instead of comparing a sample to a population. Three different tests are used to find the difference between means if the populations are assumed to be normally distributed: z test, t test, and t test for differences. The z test and the t test are used if the samples are independent. The t test for differences is used if the samples are dependent. A z test can be used to compare two proportions or two means. The null and alternate hypotheses are listed below.

$$H_0: \mu_1 \geq \mu_2 \qquad\qquad H_0: \mu_1 \leq \mu_2 \qquad\qquad H_0: \mu_1 = \mu_2$$
$$H_1: \mu_1 < \mu_2 \qquad\qquad H_1: \mu_1 > \mu_2 \qquad\qquad H_1: \mu_1 \neq \mu_2$$

z test

To use the z test for means, the samples must be independent and from normally distributed populations. Also the population standard deviations must be known or both samples must be greater than or equal to 30. The steps for testing two means are the same as for testing one sample, except the formula for the test value has changed.

The test value for a z test

$$z = \frac{(\bar{x}_1 - \bar{x}_2) - (\mu_1 - \mu_2)}{\sqrt{\frac{\sigma_1^2}{n_1} + \frac{\sigma_2^2}{n_2}}} \quad \text{or} \quad z = \frac{(\bar{x}_1 - \bar{x}_2) - (\mu_1 - \mu_2)}{\sqrt{\frac{s_1^2}{n_1} + \frac{s_2^2}{n_2}}} \quad \text{if } n_1 \text{ and } n_2 \geq 30$$

Since $\mu_1 - \mu_2$ is assumed to 0 unless otherwise stated, it will be deleted from the rest of the formulas.

t test

The t **test** can be used if the population standard deviations are not known and one or both samples are less than 30. The samples must be independent and from populations that are normally distributed. There are two formulas for the t test that depend on whether the variances can be assumed to be equal or not.

The test value for a t test when the variances are assumed to be unequal

$$t = \frac{(\bar{x}_1 - \bar{x}_2)}{\sqrt{\frac{s_1^2}{n_1} + \frac{s_2^2}{n_2}}} \qquad\qquad \text{df} = \text{smaller of } n_1 - 1 \text{ and } n_2 - 1$$

The test value for a t test when the variances are assumed to be equal

$$t = \frac{(\bar{x}_1 - \bar{x}_2)}{\sqrt{\frac{(n_1 - 1)s_1^2 + (n_2 - 1)s_2^2}{n_1 + n_2 - 2}} \sqrt{\frac{1}{n_1} + \frac{1}{n_2}}} \qquad\qquad \text{df} = n_1 + n_2 - 2$$

Dependent Samples

If the two populations are dependent, then a test for the differences will be used. Dependent samples are where a value from one sample can be paired to a value from the other sample. One sample might be the results from a post test taken after some procedure is performed. Then each individual's score taken before the procedure would be paired with the post test score. The differences are compared to see if the mean of the differences is equal to 0 or to some other specified amount. The samples have to be the same size to pair. Df $= n - 1$. The mean of the differences will be labeled μ_D for the population and $\overline{D}$ for the sample. The hypotheses are:

H_0: $\mu_D \geq 0$ H_0: $\mu_D \leq 0$ H_0: $\mu_D = 0$
H_1: $\mu_D < 0$ H_1: $\mu_D > 0$ H_1: $\mu_D \neq 0$

To find the test statistic for the mean of the differences:

1. Find the differences of the pairs of data. $D = x_1 - x_2$ (The value from the first sample subtract the value from the second sample.)
2. Find the mean of the differences. $\overline{D} = \frac{\sum D}{n}$
3. Find the standard deviation of the differences.

$$s_D = \sqrt{\frac{\sum D^2 - \frac{(\sum D)^2}{n}}{n - 1}}$$

4. Find $\frac{s_D}{\sqrt{n}}$.

5. Test statistic: $t = \frac{\overline{D} - \mu_D}{\frac{s_D}{\sqrt{n}}}$.   (NOTE: $\mu_D = 0$ unless some other value is specified. For instance, you might be testing if there is a 10 point improvement. $\mu_D = -10$)

Testing Two Sample Proportions

A z test can be used to compare proportions from two samples, if $n\hat{p}_1$, $n\hat{q}_1$, $n\hat{p}_2$ and $n\hat{q}_2$ are all greater than or equal to 5. The hypotheses are:

H_0: $p_1 \geq p_2$ H_0: $p_1 \leq p_2$ H_0: $p_1 = p_2$
H_1: $p_1 < p_2$ H_1: $p_1 > p_2$ H_1: $p_1 \neq p_2$

The test statistic for a z test for two proportions

$$z = \frac{\hat{p}_1 - \hat{p}_2}{\sqrt{\overline{pq}\left[\frac{1}{n_1} + \frac{1}{n_2}\right]}} \qquad \hat{p}_1 = \frac{x_1}{n_1} \qquad \hat{p}_2 = \frac{x_2}{n_2} \qquad \overline{p} = \frac{x_1 + x_2}{n_1 + n_2} \qquad \overline{q} = 1 - \overline{p}$$

QUESTIONS:

COMPARING 2 MEANS, INDEPENDENT

I. If both population standard deviations are known or if both n_1 and $n_2 \geq 30$.
   z test
   Test statistic:

$$z = \frac{(\bar{x}_1 - \bar{x}_2)}{\sqrt{\frac{\sigma_1^2}{n_1} + \frac{\sigma_2^2}{n_2}}}$$

   Critical value: Table F, bottom row $[(z)\ \infty]$

II. If both sample standard deviations are known and either n_1 or $n_2 < 30$.
   Variances are assumed to be unequal
   t test

$$t = \frac{(\bar{x}_1 - \bar{x}_2)}{\sqrt{\frac{s_1^2}{n_1} + \frac{s_2^2}{n_2}}}$$

   Critical value: Table F,
   df = smaller $n_1 - 1$ or $n_2 - 1$

III. If both sample standard deviations are known and either n_1 or $n_2 < 30$.
   Variances are assumed to be equal
   t test

$$t = \frac{(\bar{x}_1 - \bar{x}_2)}{\sqrt{\frac{(n_1 - 1)s_1^2 + (n_2 - 1)s_2^2}{n_1 + n_2 - 2}}\sqrt{\frac{1}{n_1} + \frac{1}{n_2}}}$$

   Critical value: Table F, df $= n_1 + n_2 - 2$

COMPARING 2 MEANS, DEPENDENT SAMPLES

I. t test

$$t = \frac{\bar{D}}{\frac{s_D}{\sqrt{n}}} \qquad D = x_1 - x_2 \qquad \bar{D} = \frac{\sum D}{n} \qquad s_D = \sqrt{\frac{\sum D^2 - \frac{(\sum D)^2}{n}}{n - 1}}$$

   Critical value: Table F, df $= n - 1$

COMPARING 2 PROPORTIONS

I. $n\hat{p}_1$, $n\hat{q}_1$, $n\hat{p}_2$ and $n\hat{q}_2 \geq 5$

   z test
   Test statistic:

$$z = \frac{\hat{p}_1 - \hat{p}_2}{\sqrt{\bar{p}\bar{q}\left[\frac{1}{n_1} + \frac{1}{n_2}\right]}}$$

$$\hat{p}_1 = \frac{x_1}{n_1} \qquad\qquad \hat{p}_2 = \frac{x_2}{n_2}$$

$$\bar{p} = \frac{x_1 + x_2}{n_1 + n_2} \qquad \bar{q} = 1 - \bar{p}$$

   Critical value: Table F, bottom row $[(z)\ \infty]$

QUESTIONS:

Checking Your Understanding

Complete this section before you do the exercises to make sure you understand the concepts. Write in the book. The answers are in the back of the book. Make a note of any questions that you wish to discuss with the instructor.

I. Tell what conditions must be met to use each of the following tests for 2 samples:
   a. z test

   b. t test

   c. t test for differences

   d. z test for proportions.

II. Select the correct answer and write the appropriate letter in the space provided.

_____ 1. If the samples are dependent, which test should be used?
   a. z test
   b. t test
   c. t test for dependence

_____ 2. The condition which must be met to use any of the tests discussed in this chapter is
   a. samples are independent.
   b. population variances are known.
   c. samples sizes must be greater than or equal to 30.
   d. populations are normally distributed.

_____ 3. If the population variances are unknown and either sample is less than 31, use the
   a. z test.
   b. t test.

___ 4. The 2 different formulas for the t test are for which two different types of problems:
   a. The samples are dependent or the samples are independent.
   b. The variances are assumed to be equal or unequal.
   c. The samples are both large or both small.
   d. The populations are normally distributed or not normally distributed.

___ 5. Unless otherwise stated, $\mu_1 - \mu_2$ is assumed to be
   a. positive.
   b. negative.
   c. zero.

QUESTIONS:

Applying Your Understanding

STUDENT: In the preceding sections, you learned the concepts of testing the difference between means and proportions and checked your understanding of the concepts. In this section, you will apply your understanding of the concepts. Study each example carefully and then try to work the following exercise. If you have any problems, see you instructor. The answers are in the back of the book.

Example 1--z test.

A high school principal claims that her students do better on a college entrance exam than another school in their area. A sample of 50 students from each high school is taken. Is the principal's claim valid at the .05 level?

Her High School
$\bar{x} = 106$
$s = 17$
$n = 50$

Other High School
$\bar{x} = 100$
$s = 16$
$n = 50$

Solution

1. State the hypotheses. The principal's claim that their score is better: $\mu_1 > \mu_2$. The opposite, $\mu_1 \leq \mu_2$.
   H_0: $\mu_1 \leq \mu_2$
   H_1: $\mu_1 > \mu_2$

2. Find the critical value. Draw and label the graph. This is a z test since both n's $\geq$ 30. one tailed, .05 level $z = 1.645$ from Table F.

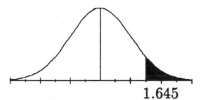

1.645

3. Compute the test value.

$$z = \frac{(\bar{x}_1 - \bar{x}_2)}{\sqrt{\frac{\sigma_1^2}{n_1} + \frac{\sigma_2^2}{n_2}}} = \frac{106 - 100}{\sqrt{\frac{17^2}{50} + \frac{16^2}{50}}} = 1.817$$

4. Reject the null.
5. The principal's claim is probably true.

Exercise 1

Is there a difference in the mean test scores for these two classes? Test at the .01 level.

Class A
$n = 35$
$\bar{x} = 78.6$
$s = 10.4$

Class B
$n = 35$
$\bar{x} = 81.2$
$s = 10.6$

Example 2--t test.

A study was conducted comparing college majors that required 12 or more hours of mathematics to majors that required less than 12 hours of mathematics. The mean income of 8 majors requiring more than 12 hours of math was $45,692 per year with a standard deviation of 2236. The mean income of 10 majors that require less than 12 hours of math was $40,416 with a standard deviation of 3524. If the variances are assumed to be unequal, test to see if there is a difference in the incomes at the .01 level.

Solution

1. State the hypotheses.
   H_0: $\mu_1 = \mu_2$
   H_1: $\mu_1 \neq \mu_2$
2. Find the critical value. Draw and label the graph. This is a t test since both sample sizes are less than 30 and the sample standard deviations are given. Use the first t formula since the variances are assumed to be unequal. two tailed, $\alpha = .01$, df = smaller $n - 1 = 8 - 1 = 7$, $t = \pm 3.499$

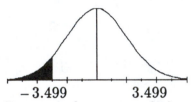

$$-3.499 \qquad 3.499$$

3. Compute the test statistic.

$$t = \frac{(\bar{x}_1 - \bar{x}_2)}{\sqrt{\frac{s_1^2}{n_1} + \frac{s_2^2}{n_2}}} = \frac{45692 - 40416}{\sqrt{\frac{2236^2}{8} + \frac{3524^2}{10}}} = 3.861$$

4. Reject the null.
5. There is a difference in the mean incomes.

Exercise 2

Two different models of cars were tested for gas mileage. Test the claim that the models have the same gas mileage at the .05 level. Assume the variances are unequal.

Car A	Car B
$n = 9$	$n = 13$
$\bar{x} = 28.6$	$\bar{x} = 32.1$
$s = 4.1$	$s = 2.1$

Example 3--t test.

Two groups were given a speed reading class using different techniques. Test the claim that Method A gives better results at the .01 level. Assume the variances are equal.

Method A	Method B
$n = 16$	$n = 12$
$\bar{x} = 44.0$	$\bar{x} = 36.5$
$s = 13.2$	$s = 10.2$

Solution

1. State the hypotheses.
    H_0: $\mu_1 \leq \mu_2$
    H_1: $\mu_1 > \mu_2$
2. Find the critical value. Draw and label a graph. This is a t test since both n's are less than 30 and the sample standard deviations are given. Use the second t formula since the variances are equal. one tailed, $\alpha = .01$, df $= n_1 + n_2 - 2 = 16 + 12 - 2 = 26$, $t = 1.706$

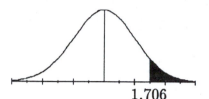

1.706

3. Compute the test statistic.

$$t = \frac{(\bar{x}_1 - \bar{x}_2)}{\sqrt{\dfrac{(n_1 - 1)s_1{}^2 + (n_2 - 1)s_2{}^2}{n_1 + n_2 - 2}} \sqrt{\dfrac{1}{n_1} + \dfrac{1}{n_2}}}$$

$$= \frac{44.0 - 36.5}{\sqrt{\dfrac{(16 - 1)13.2^2 + (12 - 1)10.2^2}{26}} \sqrt{\dfrac{1}{16} + \dfrac{1}{12}}} = 1.634$$

4. Accept the null.
5. There is not enough evidence to claim that Method A is better.

Exercise 3

Machine A and Machine B fill bags of grain. Does Machine A put less grain in than Machine B at the .01 level? Assume the variances are equal.

Machine A	Machine B
$n = 18$	$n = 14$
$\bar{x} = 48.3$	$\bar{x} = 50.4$
$s = 1.8$	$s = 1.8$

Example 4--*t* test of differences.

A drug is used to lower blood pressure. A sample of eight patients were given the drug. Their blood pressure was taken before and after the drug. Test the claim, $\alpha = .05$, that the drug lowers the pressure.

Before	After
135	124
120	118
155	148
137	128
115	112
187	172
173	161
160	151

Solution

The samples are dependent since the blood pressure after the drug is related to the blood pressure before the drug, so the t test for differences must be used.

1. State the hypotheses. If the drug lowers the blood pressure, the second value will be less, so $\mu_1 - \mu_2$ would be positive, and μ_D would be positive.
   H_0: $\mu_D \le 0$
   H_1: $\mu_D > 0$

2. Find the critical value. Draw and label a graph.
   one tailed test, $\alpha = .05$, df $= n - 1 = 8 - 1 = 7$, $t = 1.895$

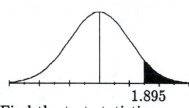

   1.895

3. Find the test statistic.
   a. Find $x_1 - x_2 = D$ and D^2.

x_1	x_2		D	D^2
135	124		11	121
120	118		2	4
155	148		7	49
137	128		9	81
115	112		3	9
187	172		15	225
173	161		12	144
160	151		9	81
			68	714

   b. $\overline{D} = \dfrac{\sum D}{n} = \dfrac{68}{8} = 8.5$

   c. $s_D = \sqrt{\dfrac{\sum D^2 - \dfrac{(\sum D)^2}{n}}{n-1}} = \sqrt{\dfrac{714 - \dfrac{68^2}{8}}{7}} = 4.408$

   d. $\dfrac{s_D}{\sqrt{n}} = \dfrac{4.408}{\sqrt{8}} = 1.558$

   e. $t = \dfrac{\overline{D}}{\dfrac{s_D}{\sqrt{n}}} = \dfrac{8.5}{1.558} = 5.456$

4. Reject the null.
5. The drug does lower blood pressure.

Exercise 4

IQ tests were administered to 10 students at 8 a.m. Two months later, the same test was administered to the same subjects at 4 p.m. Test at the .01 level to see if there is a difference in the performances.

8 a.m.	105	104	112	102	124	118	114	108	113	96
4 p.m.	99	98	110	105	124	116	112	105	114	94

Example 5--Testing proportions.

A study was conducted to compare the proportion of smokers under 20 to the proportion of middle aged smokers. A sample of 200 people under 20 had 56 smokers. A sample of 175 between the ages of 35 and 50 showed 55 smokers. Is there a lower proportion in the younger group? Test at the .01 level.

Solution

To see if the z test can be used, $n\hat{p}_1$, $n\hat{q}_1$, $n\hat{p}_2$ and $n\hat{q}_2 \geq 5$.

$\hat{p}_1 = \frac{x_1}{n_1} = \frac{56}{200} = .28 \qquad \hat{q}_1 = 1 - \hat{p}_1 = 1 - .28 = .72$

$n\hat{p}_1 = 200(.28) = 56 \qquad n\hat{q}_1 = 200(.72) = 144$

$\hat{p}_2 = \frac{x_2}{n_2} = \frac{55}{175} = .314 \qquad \hat{q}_2 = 1 - \hat{p}_2 = 1 - .314 = .686$

$n\hat{p}_1 = 175(.314) = 55 \qquad n\hat{q}_1 = 175(.686) = 120 \qquad$ All are ≥ 5, so can use the z test.

1. State the hypotheses.
   H_0: $p_1 \geq p_2$
   H_1: $p_1 < p_2$

2. Find the critical value. Draw and label the graph. one tail, $\alpha = .01$, $z = 2.326$

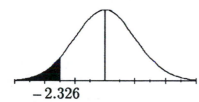

-2.326

3. Find the test statistic.

$z = \dfrac{\hat{p}_1 - \hat{p}_2}{\sqrt{\bar{p}\bar{q}\left[\frac{1}{n_1} + \frac{1}{n_2}\right]}}$
$\qquad \hat{p}_1 = \frac{x_1}{n_1} = \frac{56}{200} = .28 \qquad \hat{p}_2 = \frac{x_2}{n_2} = \frac{55}{175} = .314$

$\qquad \bar{p} = \dfrac{x_1 + x_2}{n_1 + n_2} = \dfrac{56 + 55}{200 + 175} = \dfrac{111}{375} = .296$

$\qquad \bar{q} = 1 - \bar{p} = 1 - .296 = .704$

$z = \dfrac{.28 - .314}{\sqrt{(.296)(.704)\left[\frac{1}{200} + \frac{1}{175}\right]}} = -.7196$

4. Accept the null.
5. Not enough evidence to accept that the younger group has fewer smokers.

Exercise 5

A study was done comparing recycling done in large cities to recycling done in smaller towns. 250 people in a city with a population of over 250,000 were interviewed and 115 said they did some type of recycling. Out of 120 people in a town of under 75,000 population, 41 recycled. Is there a difference in the proportions at the .05 level?

QUESTIONS:

Practice Test

1. Brand X and Brand Y of pain relievers were tested to see how much ibuprofen each tablet contained. Does Brand X have more ibuprofen? Test at the .05 level.

Brand X	Brand Y
$n = 36$	$n = 35$
$\bar{x} = 358$	$\bar{x} = 345$
$s = 10$	$s = 14$

2. A bank studied the mean time clerks spent with a customer. A sample of 12 male clerks spent a mean time of 1.4 minutes with each customer with $s = .31$. A sample of 14 female clerks spent a mean time of 1.7 minutes with $s = .16$. Assume the variances are unequal and test at the .01 level the claim that men and women clerks spend the same amount of time with each customer.

3. A study was conducted on pregnant women who smoked and those who did not smoke, to see if maternal smoking had an effect on birthweight. At the .05 level, test the claim that maternal smoking can lower the birthweight of the child. Assume the variances to be equal.

Smokers	Non-Smokers
$n_1 = 20$	$n_2 = 20$
$\bar{x}_1 = 6.9$	$\bar{x}_2 = 7.2$
$s_1 = .5$	$s_2 = .4$

4. Students were paired by matching their IQ's and grades in previous mathematics courses taken. Group A attended lecture and did homework. Group B watched videotapes and did work on a computer. Test to see if there is a significant difference in the final exam scores for the two groups at the .05 level.

Lecture	Tapes
95	99
87	91
91	88
85	90
81	87
79	78
74	79
73	81
71	65
69	74

5. In a class of remedial English that was taught by the lecture method, 52 out of 75 students completed the course with a passing grade. In the same course with computer assisted instruction, 69 out of 95 completed the course with a passing grade. At the .05 level, is there a significant difference in the passing rates for the two different methods of instruction?

CHAPTER 11
CORRELATION AND REGRESSION
Understanding Correlation and Regression

Introduction

Correlation analysis is used to determine if there is a relationship, or correlation, between two variables and what is the strength of the correlation. The relationship can be positive or negative. Two variables have a **positive correlation** if when one variable increases, so does the other. If one variable increases and the other decreases, they have a **negative correlation**. **Regression analysis** is used to determine what type of relationship exists and to make predictions using the relationship.

Simple Correlation

In a **simple correlation**, only two variables are studied at once. The two variables are the independent and the dependent variable. The **independent variable** is the variable that can be controlled or picked. The **dependent variable** is the variable that you assume depends on the other variable. The independent variable will be used to predict the dependent variable if there is a correlation between the two variables. The independent variable is labeled X and the dependent variable is labeled Y.

A **scatter plot** is a graph with the independent variable plotted along the bottom and the dependent variable along the side. A point is placed on the graph for each pair of numbers, but the points are not connected with a line. The scatter plot helps you to see if there is a linear correlation. The closer the points are to being a straight line, the more linear correlation there is between the two variables. If the line has a positive slope (line is going up as you move from left to right), then there is a positive correlation. If the slope is negative, the correlation is negative.

The **correlation coefficient** measures the strength of the correlation between the two variables. The correlation, called r, is a value between -1.00 and $+1.00$. -1.00 would indicate a perfect negative correlation and the scatter plot would have all the points in a straight line with the slope down. $+1.00$ would indicate a perfect positive correlation and the points would be in a straight line going up. The closer the correlation is to 0, the less the correlation there is between the two variables.

$$r = \frac{n(\sum XY) - (\sum X)(\sum Y)}{\sqrt{\left[n\sum X^2 - (\sum X)^2\right]\left[n\sum Y^2 - (\sum Y)^2\right]}} \qquad n = \textbf{pairs of numbers}$$

The **coefficient of determination** tells how much variation in the dependent variable is explained by the independent variable. The coefficient is r^2 and is usually expressed as a percent. $1 - r^2$ is the **coefficient of non-determination** and is the variation in the dependent variable that is not explained by the independent variable. The correlation coefficient can be tested for significance by using a t test and following the procedures for hypothesis testing, or by comparing the r to a value from Table I. The null hypothesis is that there is no correlation, or that $r = 0$. The alternate hypothesis is that there is a correlation, or $r \neq 0$.

To test r:

1. Find the value from Table I for the desired level of significance. $df = n - 2$
2. If r is between $-$Table and $+$Table value, accept the null hypothesis. The correlation could

be 0.

3. If r is smaller than $-$Table value or greater than $+$Table value, reject the null. There is a correlation.

Regression Analysis

The **regression line** is the equation of the line that best fits the points of the scatter plot if there is a correlation. The regression line is written $y' = a + bx$ where a is the y intercept and b is the slope of the line. y' will be the predicted value of y for any given x value. If the scatter plot indicates the line is going up, the slope b will be positive and r will be positive. If the line is going down, b and r will be negative.

The regression line: $y' = a + bx$

$$a = \frac{\sum Y(\sum X^2) - \sum X(\sum XY)}{n(\sum X^2) - (\sum X)^2} \qquad b = \frac{n(\sum XY) - (\sum X)(\sum Y)}{n\sum X^2 - (\sum X)^2}$$

The **standard error estimate**, s_{est}, is the standard deviation of the observed y values about the predicted y' values, or an average of how much error there will be in each predicted y.

$$s_{est} = \sqrt{\frac{\sum(y - y')^2}{n - 2}} \qquad \text{or} \qquad s_{est} = \sqrt{\frac{\sum Y^2 - a\sum Y - b\sum XY}{n - 2}}$$

A **confidence interval** for a predicted y can be found using the standard error of estimate if the sample size is larger than 100. Instead of predicting a single value for y, you would be more confident in saying that y would be between 2 values for a given x.

A 95% confidence interval, if $n > 100$

$$y' - 1.96s_{est} \le y \le y' + 1.96s_{est}$$

where y' is the value obtained from $y' = a + bx$ for a given x.

NOTES:

Checking Your Understanding

Complete this section before you do the exercises to make sure you understand the concepts. Write in the book. The answers are in the back of the book. Make a note of any questions that you wish to discuss with the instructor.

I. Tell if each of the following scatter plots would have a positive or negative correlation and indicate if there is a strong correlation, a fair correlation, or almost no correlation.

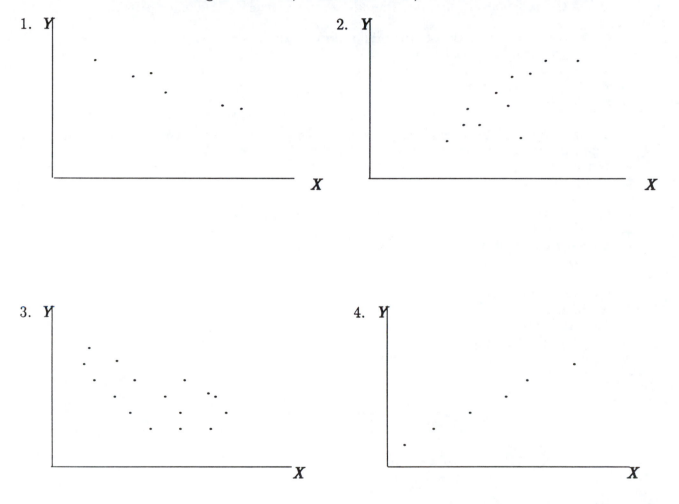

II. Indicate which would be the independent variable and which would be the dependent variable.

1. Score on the first test   Final grade

2. Dexterity   Number of alcoholic drinks

3. Gas mileage   Car speed

4. High school grade point average   Freshman college grade point average

5. Grade in college   Age

6. Crop yield   Rainfall

III. Select the correct answer and write the appropriate letter in the space provided.

_____ 1. If you compared a group of students' scores on the first test in a class to their final grade, you might expect
   a. a positive correlation.
   b. a negative correlation.
   c. no correlation.

_____ 2. If you compared the speed a car was driven to its gas mileage, you would probably get
   a. a positive correlation.
   b. a negative correlation.
   c. no correlation.

_____ 3. The variation of the dependent variable that is explained by the independent variable is
   a. the correlation coefficient.
   b. the coefficient of determination.
   c. the coefficient of non-determination.

_____ 4. The correlation coefficient can be between
   a. 0 and 1.00.
   b. 1.00 and 100.
   c. -1.00 and 1.00.
   d. 1 and 10.

QUESTIONS:

Applying Your Understanding

STUDENT: In the preceding sections, you learned the concepts of correlation and regression and checked your understanding. In this section, you will apply your understanding of the concepts. Study each example carefully and then try to work the following exercise. If you have any problems, see your instructor. The answers are in the back of the book.

Example 1--Scatter plot.

Make a scatter plot for the following data. Interpret.

Grade on first test	Final grade
73	70
86	80
93	96
92	85
72	68
65	68
58	62
75	78

Solution

Decide which is the dependent and the independent variable. The final grade depends on the first test, so the final grade would be Y and first test X. Draw a graph and label numbers from 50 to 100 along the bottom and the side. Find 73 (the first X value) along the bottom and 70 along the side (the first Y value) and put a point.

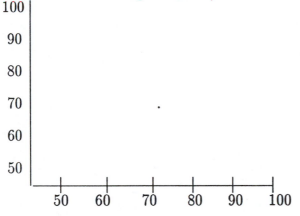

Put a point on the graph for each of the other pairs of values. Label.

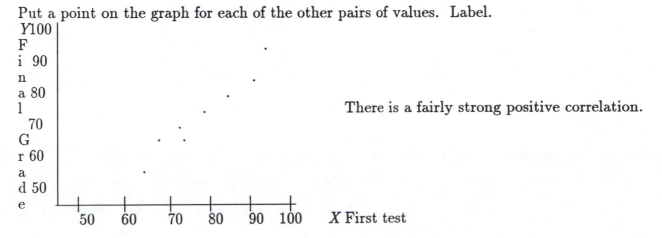

There is a fairly strong positive correlation.

Exercise 1

Draw a scatter plot for the following data.

Number of absences	Final grade
0	86
1	91
2	78
2	83
3	75
3	62
4	70
5	68
6	56

Example 2--Correlation coefficient.

Find the correlation coefficient and test the significance of r at the .05 level for the data in Example 1.

Solution

$$r = \frac{n(\sum XY) - (\sum X)(\sum Y)}{\sqrt{\left[n\sum X^2 - (\sum X)^2\right]\left[n\sum Y^2 - (\sum Y)^2\right]}}$$

n = pairs of numbers

Make columns X, Y, X^2, Y^2, XY and total the columns.

X	Y	X^2	Y^2	XY
73	70	5329	4900	5110
86	80	7396	6400	6880
93	96	8649	9216	8928
92	85	8464	7225	7820
72	68	5184	4624	4896
65	68	4225	4624	4420
58	62	3364	3844	3596
75	78	5625	6084	5850
614	607	48,236	46,917	47,500

$$r = \frac{8(47,500) - 614(607)}{\sqrt{[8(48,236) - 614^2][8(46,917) - 607^2]}} = .933$$

Test r:

Table I, .05 level, df $= n - 2 = 6$, Table value $= .707$
.933 is not between $-.707$ and $+.707$, so there is a correlation.
.933 is a very strong positive correlation.

Exercise 2

Find the correlation coefficient and test the significance of r at the .01 level for the data in Exercise 1.

Example 3--Coefficients of Determination and Non-determination.

Find the coefficients of determination and non-determination for Example 2. Interpret.

Solution

coefficient of determination $= r^2 = .933^2 = .87$

87% of the variations in final grades can be determined by the variations in the first test.

coefficient of non-determination $= 1 - r^2 = 1 - .87 = .13$

13% of the variations in final grades can not be determined by the variations in the first test.

> **Exercise 3**
>
> Find the coefficients of determination and non-determination for Exercise 2. Interpret.

Example 4--Regression line.

Find the equation for the regression line for Example 1.

Solution

$y' = a + bx$

$$a = \frac{\sum Y(\sum X^2) - \sum X(\sum XY)}{n(\sum X^2) - (\sum X)^2} \qquad b = \frac{n(\sum XY) - (\sum X)(\sum Y)}{n\sum X^2 - (\sum X)^2}$$

$$a = \frac{607(48,236) - 614(47,500)}{8(48,236) - 614^2} \qquad b = \frac{8(47,500) - 614(607)}{8(48,236) - 614^2}$$

$a = 12.85 \qquad\qquad\qquad b = .82$

$y' = 12.85 + .82x$

> **Exercise 4**
>
> Find the regression equation for Exercise 1.

Example 5--Standard error of estimate.

Find the standard error of estimate for Example 1.

Solution

$$s_{est} = \sqrt{\frac{\sum Y^2 - a\sum Y - b\sum XY}{n - 2}} = \sqrt{\frac{46,917 - 12.85(607) - .82(47,500)}{6}} = 5.28$$

Each predicted final grade will have an error of about 5.28 points.

> **Exercise 5**
>
> Find the standard error of estimate for Exercise 1.

Example 6--Confidence Interval.

For a sample of 180, the regression line is $y' = 25 + 4x$ and the standard error of estimate is 3. Find a 95% confidence interval for $x = 12$.

Solution

Find the predicted value for y. $y' = 25 + 4x = 25 + 4(12) = 73$

$y' - 1.96s_{est} \leq y \leq y' + 1.96s_{est}$

$73 - 1.96(3) \leq y \leq 73 + 1.96(3)$
$73 - 5.88 \leq y \leq 73 + 5.88$
$67.12 \leq y \leq 78.88$

You can be 95% confident that for a x of 12, y will be between 67.12 and 78.88.

> **Exercise 6**
>
> For a sample of 120, the regression equation is $y' = 18 - .4x$ and the standard error of estimate is 1.3. Find a 95% confidence interval for $x = 6$.

NOTES:

Practice Test

1. Number of alcoholic drinks   Score on a dexterity test

Number of alcoholic drinks	Score on a dexterity test
2	15
1	18
3	11
4	7
2	10
1	16
5	5
6	2

  a. Draw a scatter plot.

  b. Find the coefficient of correlation and test the significance at the .05 level.

  c. Find the coefficients of determination and non-determination.

d. Find the regression equation.

e. Find the standard error of estimate.

2. For a sample of 200, the regression line is $y' = 975 - 73x$ and the standard error of estimate is 45. Find the 95% confidence interval for $x = 33$.

CHAPTER 12
CHI-SQUARE
Understanding Chi-Square

Introduction

The **chi-square** test is used to test hypotheses concerning variances, frequency distributions, or the independence of two variables. Chi-square is written χ^2. The χ^2 test is conducted like the t and z tests explained in earlier chapters, except a χ^2 value is never negative even for a left tail test. The χ^2 distribution is not symmetrical and the graph is not bell shaped.

Steps for the χ^2 test:
1. State the hypotheses.
2. Find the critical value or values and draw the graph.
3. Compute the test value.
4. Make the decision.
5. Summarize the results.

The table value for a χ^2 test is found in Table G. (Tables are in the back of the text and this manual.) There are three ways to find a table value depending on what the hypotheses are. The degrees of freedom is $n - 1$ where n is the sample size.

Table values for χ^2:
1. One tailed right
   H_0: $\leq$
   H_1: $>$

   Find the desired α at the top of the table and the degrees of freedom at the far left column to find the table value. Reject the null if the test value is greater than the table value.

table value

2. One tailed left
   H_0: $\geq$
   H_1: $<$

   Find $1 - \alpha$. Look for $1 - \alpha$ at the top of the table and the degrees of freedom to find the table value. Reject the null if the test value is less than the table value.

table value

3. Two tailed test
   H_0: $=$
   H_1: $\neq$

   Find $\frac{\alpha}{2}$. The right table value is $\frac{\alpha}{2}$ for the desired degrees of freedom. The left table value is $1 - \frac{\alpha}{2}$ for the desired degrees of freedom. Reject the null if the test value is less than the left table value or greater than the right table value.

left table value   **right table value**

Test for a Single Variance

Tests for variances are used in situations where consistency is important. Claims about a single variance compare a sample variance, s^2, to the population variance, σ^2. To use the χ^2, the sample must be randomly selected from a normal population and the observations must be independent. (Remember the variance is the standard deviation squared. If the standard deviation is given, you must square the value to get the variance.)

χ^2 **formula to test a single variance:**

$$\chi^2 = \frac{(n - 1)\, s^2}{\sigma^2}$$

df $= n - 1$ n = sample size
s^2 = sample variance
σ^2 = population variance

A confidence interval for a population variance can be determined from a sample variance using the χ^2 value.

Confidence Interval for a Population Variance

$$\frac{(n - 1)\, s^2}{\chi^2_{larger}} < \sigma^2 < \frac{(n - 1)\, s^2}{\chi^2_{smaller}}$$

n = sample size χ^2_{larger} = table value for $\frac{\alpha}{2}$
s^2 = sample variance
df $= n - 1$ $\chi^2_{smaller}$ = table value for $1 - \frac{\alpha}{2}$

Confidence Interval for a Population Standard Deviation

$$\sqrt{\frac{(n - 1)\, s^2}{\chi^2_{larger}}} < \sigma < \sqrt{\frac{(n - 1)\, s^2}{\chi^2_{smaller}}}$$

Test for Goodness of Fit

The **test for goodness of fit** is used to see if a frequency distribution fits a specific pattern. The given number in each group is called the **observed frequency** and the frequency for the specific pattern is called the **expected frequency**. The expected frequencies should be at least 5 for each group. If an expected value for any class is less than 5, than that class should be combined with another class so that each expected value is 5 or more. This is always a one tailed right test, so the table value is α for df = number categories $-$ 1. The hypotheses are:

H_0: **There is a fit.**
H_1: **There is not a fit.**

table value

Reject the null if the computed value is larger than the table value.

Formula for χ^2 Goodness of Fit

$$\chi^2 = \sum \frac{(O - E)^2}{E}$$

O = observed value
E = expected value
df = number of categories − 1

Tests for Independence

Tests for independence are used to see if there is a difference in frequency distributions for two different groups. The null hypothesis is that the frequency distributions are independent of which group they are in. The alternate hypothesis is that the frequencies depend on which group they are in. This is always a one tailed right test. The two frequencies are listed in a contingency table, using as many row and columns as needed.

	Column 1	Column 2	Column 3	
Row 1	$Cell_{1,1}$	$Cell_{1,2}$	$Cell_{1,3}$	Total of Row 1
Row 2	$Cell_{2,1}$	$Cell_{2,2}$	$Cell_{2,3}$	Total of Row 2
	Total of Column 1	Total of Column 2	Total of Column 3	Grand Total

Each block is called a cell and is numbered by giving the row and column that the cell is in. $Cell_{1,2}$ would be row 1 and column 2. The row number is always given first. The expected value for each cell is found by multiplying the row sum and the column sum for that cell and dividing by the grand total. The grand total is the total of the column totals and is also equal to the total of the row totals. The table value is the value for the given α and df = (R − 1)(C − 1). R is the number of rows and C is the number of columns.

Formula for Test Value for Test for Independence:

$$\chi^2 = \sum \frac{(O - E)^2}{E}$$

O = observed value for each cell
E = expected value for each cell = $\dfrac{(Row\ Sum)(Column\ Sum)}{Grand\ Total}$

df = (R − 1)(C − 1)

R = number of rows
C = number of columns

NOTES:

NOTES:

Checking Your Understanding

Complete this section before you do the exercises to make sure you understand the concepts. Write in the book. The answers are in the back of the book. Make a note of any questions that you wish to discuss with the instructor.

I. List the conditions and uses for χ^2 test for a single variance.

II. List the conditions and uses for χ^2 test for goodness of fit.

III. List the conditions and uses for the χ^2 test for independence.

QUESTIONS:

Applying Your Understanding

STUDENT: In the preceding sections, you learned the concepts of the chi-square tests and checked your understanding of the concepts. In this section, you will apply your understanding of the concepts. Study each example carefully and then try to work the following exercise. If you have any problems, see your instructor. The answers to the exercises are in the back of the book.

Example 1--Single variance.

A manufacturer wants to know if the variance of the size of the diameter of a nut is equal to 12. A sample of 17 nuts had a variance of 10.6. Test if the variance is 12 at the .01 level.

Solution

1. State the hypotheses.
   H_0: $\sigma^2 = 12$
   H_1: $\sigma^2 \neq 12$

2. Find the critical values and draw a graph.
   To find the critical values, first find $\frac{\alpha}{2}$ since this is a two tail test. $\frac{\alpha}{2} = \frac{.01}{2} = .005$
   df $= n - 1 = 17 - 1 = 16$
   To find the left table value, look at $1 - \frac{\alpha}{2} = 1 - .005 = .995$ and df = 16.
   left table value = 5.142
   To find the right table value, look at $\frac{\alpha}{2} = .005$ and df = 16.
   right table value = 34.267

   5.142   34.267

3. Compute the test value.
   $$\chi^2 = \frac{(n-1)\,s^2}{\sigma^2} = \frac{(17-1)10.6}{12} = 14.133$$

4. Accept the null.

5. The variance is 12.

Exercise 1
Test the claim at the .05 level that the variance is 84 if a sample of 28 had a variance of 79.

Example 2--Single variance.

A teacher claims that the variance on a certain test is less than 68. A sample of 58 students had a standard deviation of 5.8. Test the teacher's claim at the .01 level.

Solution

1. State the hypotheses.
   H_0: $\sigma^2 \geq 68$
   H_1: $\sigma^2 < 68$

2. Find the critical value and draw a graph.
   The table value for a one tail left test uses $1 - \alpha = 1 - .01 = .99$, df $= n - 1 = 58 - 1 = 57$. Since 57 is not on the table, take the closest value that is **lower** than 57, table value for .99, and df $= 50$ is 29.707.

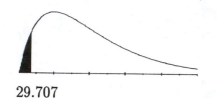

29.707

3. Compute the test value. Since the standard deviation is given, remember to square the standard deviation to find the variance.
   $$\chi^2 = \frac{(n - 1)\,s^2}{\sigma^2} = \frac{(58 - 1)5.8^2}{68} = 28.198$$

4. Reject the null.

5. The variance is less than 68.

Exercise 2

Test the claim at the .01 level that the population variance is less than 125 if a sample of 43 had a standard deviation of 8.9.

Example 3--Confidence interval for a variance.

Find the 95% confidence interval for a population variance if a sample of 28 had variance of 116.

Solution

$$\frac{(n - 1)\,s^2}{\chi^2_{larger}} < \sigma^2 < \frac{(n - 1)\,s^2}{\chi^2_{smaller}}$$

Confidence intervals are two tailed so first find $\frac{\alpha}{2} = \frac{.05}{2} = .025$, df $= n - 1 = 28 - 1 = 27$.

χ^2_{larger} = table value for $\frac{\alpha}{2}$ and df $= 27$ $\qquad$ $\chi^2_{larger} = 43.194$

$\chi^2_{smaller}$ = table value for $1 - \frac{\alpha}{2} = 1 - .025 = .975$, df $= 27$ $\qquad$ $\chi^2_{smaller} = 14.573$

$$\frac{(28 - 1)\,116}{43.194} < \sigma^2 < \frac{(28 - 1)\,116}{14.573}$$

$72.510 < \sigma^2 < 214.918$

Exercise 3

Find the 99% confidence interval for the population if a sample of 19 had a variance of 83.

Example 4--Goodness of fit.

A store sold 12 stereos on Monday, 17 on Tuesday, 28 on Wednesday, 17 on Thursday, and 26 on Friday. At the .01 level, test if there is a difference in the number of stereos sold on each weekday.

Solution

1. State the hypotheses.
   H_0: There is no difference in the number of stereos sold on each weekday.
   H_1: There is a difference in the number of stereos sold on each weekday.

2. Find the critical value and draw the graph.
   Goodness of fit is always a one tailed right test, df = number of categories $- 1 = 5 - 1$ $= 4$. Table value for .01 and df $= 4$ is 13.277.

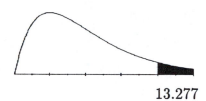

13.277

3. Compute the test value.

$$\chi^2 = \Sigma \frac{(O - E)^2}{E}$$

Since there are 100 frequencies (sum of all the stereos sold), if there were no difference in the number sold on each day, the expected number of stereos sold on each weekday would be $\frac{100}{5} = 20$.

O	E	O − E	$(O - E)^2$	$\frac{(O - E)^2}{E}$
12	20	− 8	64	$\frac{64}{20} = 3.20$
17	20	− 3	9	$\frac{9}{20} = 0.45$
28	20	8	64	$\frac{64}{20} = 3.20$
17	20	− 3	9	$\frac{9}{20} = 0.45$
26	20	6	36	$\frac{36}{20} = \underline{1.80}$
				9.10

$\chi^2 = 9.10$

4. Accept the null.

5. There is no significant difference in the number of stereos sold on each weekday.

Exercise 4

A company had 36 absences on Monday, 26 on Tuesday, 10 on Wednesday, 20 on Thursday, and 28 on Friday. At the .05 level, is there difference in the number of absences per day?

Example 5--Goodness of fit.

A store manager believes that 20% of the people who come into the store spend less than $10, 20% spend between $11 and $30, 50% spend between $31, and $50 and 10% spend more than $51. If during one morning, 40 people spent less than $10, 25 spent between $11 and $30, 90 spent between $31 and $50, and 5 spent more than $51, test the manager's claim at the .05 level.

Solution

1. State the hypotheses.
    H_0: There is 20% in the first group, 20% in the second, 50% in the third and 10% in the last.
    H_1: There is not.

2. Find the table value and draw the graph.
    df = number of categories $- 1 = 4 - 1 = 3$, one tail right test, $\alpha = .05$
    table value = 7.815

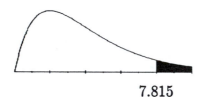

7.815

3. Compute the test value.
    There are a total of 160 frequencies. If the first and the second group had 20%, E = 160(.2) = 32. For the third group, E = 160(.5) = 80. For the fourth group, E = 160(.1) = 16.

O	E	O – E	$(O - E)^2$	$\dfrac{(O - E)^2}{E}$
40	32	8	64	$\dfrac{64}{32} = 2.00$
25	32	-7	49	$\dfrac{49}{32} = 1.53$
90	80	10	100	$\dfrac{100}{80} = 1.25$
5	16	-11	121	$\dfrac{121}{16} = \underline{7.56}$
				12.34

$\chi^2 = 12.34$

4. Reject the null.

5. The amounts spent do not follow the pattern the manager believes.

Exercise 5

The head of a mathematics department believes that 65% of the students enroll in morning classes, 25% in afternoon classes, and 10% in night classes. If one semester, 155 students enrolled in morning classes, 30 in the afternoon classes, and 15 in night classes, test the claim at the .05 level.

Example 6--Independence.

A company wants to know if the number of defective parts made is dependent on the day of the week. Test for independence at the .05 level.

	Monday	Tuesday	Wednesday	Thursday	Friday
Non-defective	220	250	260	240	250
Defective	40	25	18	20	38

Solution

1. State the hypotheses.
   H_0: The number of defectives is independent of the day of the week.
   H_1: The number of defectives is dependent on the day of the week.

2. Find the critical value and draw a graph.
   number of rows = R = 2, number of columns = C = 5
   df = (R − 1)(C − 1) = (2 − 1)(5 − 1) = (1)(4) = 4
   right tail, .05, table value = 9.488

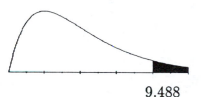

9.488

3. Compute the test value.
   Total each row and each column on the contingency table and find the grand total. The grand total is the total of the column totals and is also the total the row total.

	C_1	C_2	C_3	C_4	C_5	
R_1	220	250	260	240	250	1220
R_2	40	25	18	20	38	141
	260	275	278	260	288	1361

Find the expected value for each cell. $E = \dfrac{(\text{Row sum})(\text{Column sum})}{\text{Grand total}}$

$$C_{1,1} = \frac{(1220)(260)}{1361} = 233 \quad \text{(Cell from row 1, column 1)}$$

$$C_{1,2} = \frac{(1220)(275)}{1361} = 247 \quad \text{(Cell from row 1, column 2)}$$

$$C_{1,3} = \frac{(1220)(278)}{1361} = 249 \quad \text{(Cell from row 1, column 3)}$$

$$C_{1,4} = \frac{(1220)(260)}{1361} = 233$$

$$C_{1,5} = \frac{(1220)(288)}{1361} = 258$$

$$C_{2,1} = \frac{(141)(260)}{1361} = 27$$

$$C_{2,2} = \frac{(141)(275)}{1361} = 28$$

$$C_{2,3} = \frac{(141)(278)}{1361} = 29$$

$$C_{2,4} = \frac{(141)(260)}{1361} = 27$$

$$C_{2,5} = \frac{(141)(288)}{1361} = 30$$

List the observed value from each cell (O) and expected value from each cell (E). Find $(O - E)$, $(O - E)^2$, and $\frac{(O - E)^2}{E}$ for each cell.

$$\chi^2 = \sum \frac{(O - E)^2}{E}$$

O	E	O – E	$(O - E)^2$	$\frac{(O - E)^2}{E}$
220	233	– 13	169	$\frac{169}{233}$ = .73
250	247	3	9	$\frac{9}{247}$ = .04
260	249	11	121	$\frac{121}{249}$ = .49
240	233	7	49	$\frac{49}{233}$ = .21
250	258	– 8	64	$\frac{64}{258}$ = .25
40	27	13	169	$\frac{169}{27}$ = 6.26
25	28	– 3	9	$\frac{9}{28}$ = .32
18	29	– 11	121	$\frac{121}{29}$ = 4.17
20	27	– 7	49	$\frac{49}{27}$ = 1.81
38	30	8	64	$\frac{64}{30}$ = 2.13
				16.41

$$\chi^2 = 16.41$$

4. Reject the null.

5. The number of defectives made is dependent on the day of the week.

Exercise 6

A study was made of men and women who smoked. Is the number of smokers dependent on sex at the .01 level?

	Smoke	Don't Smoke
Male	60	100
Female	40	80

Practice Test

1. Test at the .01 level the claim that the population variance is less than 158 if a sample of 10 had a variance of 164.

2. Find the 90% confidence interval for the population variance if a sample of 12 had a variance of 360.

3. An instant oatmeal mix is considering adding flavors to its mix. 200 people tested the flavors and their preferences. Is there a preference for the flavor at the .05 level?
   Plain   20
   Cinnamon   58
   Apple   48
   Maple   22
   Peach   52

4. Is the grade dependent on sex? Test at the .05 level.

	A	B	C	D or lower
Males	8	17	25	10
Females	12	10	21	7

CHAPTER 13
THE F TEST AND ANALYSIS OF VARIANCE
Understanding the F Test and Analysis of Variance

Introduction

The **F test** is used to compare two sample variances or to test claims about three or more means. The F test can only indicate if there is a difference among the means, not where the difference is. If there is a difference, the **Scheffè test** or the **Tukey test** is used to find where the difference is.

Two Variances

When the F test is used to compare two variances, the samples must be independent and from normally distributed populations. The group with the larger variance is always labeled with a subscript "1." The larger sample variance is labeled s_1^2, the sample size in the sample with the larger variance is n_1. Since the larger variance is labeled sub 1, the only one tailed test that can be used is the right tailed. There are only two possible hypotheses for the F test of two variances.

one tailed right
H_0: $\sigma_1^2 \leq \sigma_2^2$
H_1: $\sigma_1^2 > \sigma_2^2$

two tailed
H_0: $\sigma_1^2 = \sigma_2^2$
H_1: $\sigma_1^2 \neq \sigma_2^2$

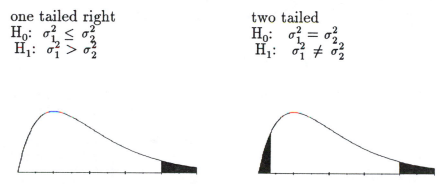

For both tests, only the right tailed value is given, since $s_1^2 > s_2^2$ because the labels are picked that way. Table H is used to find the table value. NOTE: there is a different table for each α.

Important Points for F Test

1. $F = \dfrac{s_1^2}{s_2^2}$ where s_1^2 is the larger variance and is always placed in the numerator.

2. If conducting a one tailed test, find α on the table. If conducting a two tailed test, find $\frac{\alpha}{2}$ on the table. Make a right tailed graph for either test.

3. To use Table H:
    dfN = degrees of freedom of the numerator = $n_1 - 1$
    dfD = degrees of freedom of the denominator = $n_2 - 1$
    NOTE: n_1 is the sample size for the sample with the larger variance, but will not necessarily be the larger sample size.

4. If the standard deviations are given, square them to get the variances.

5. When the degrees of freedom are not listed on the table, the closest smaller value should be used.

Steps for F Test for Variances:
1. State the hypotheses.
    H_0: $\sigma_1^2 \leq \sigma_2^2$
    H_1: $\sigma_1^2 > \sigma_2^2$

    H_0: $\sigma_1^2 = \sigma_2^2$
    H_1: $\sigma_1^2 \neq \sigma_2^2$

2. **Find the table value and draw a graph.**
   dfN = n_1 − 1
   dfD = n_2 − 1
   For one tailed test, use α. For two tailed test, use $\frac{\alpha}{2}$.

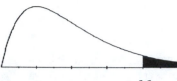

   table value for both one and two tailed test

3. **Find the test value.** $F = \dfrac{s_1^2}{s_2^2}$ where s_1^2 is the larger variance.

4. **Make a decision.** Reject the null if the test value is larger than the table value.
5. **Summarize the results.**

Analysis of Variance

The F test can be used to compare three or more means if the samples are from normally distributed populations, the samples are independent, and the variances of the populations are equal. The **analysis of variance**, or **ANOVA**, is used to test the difference between three or more means. This is always a one tailed right test. The table value comes from Table H for the desired α and for:
   dfN = k − 1 where k is the number of groups
   dfD = N − k where N is the sum of all the sample sizes

The null hypothesis is that all means are equal, and the alternate is that at least one mean is different from the others. The ANOVA does not show which means are different, just that there is a difference in at least one.

Test value for ANOVA:

$$F = \frac{s_B^2}{s_W^2} \qquad \text{where}$$

$$s_B^2 = \frac{\Sigma n(\bar{x} - \bar{x}_{GM})}{k - 1} \qquad \bar{x} = \text{mean of each group} \qquad n = \text{sample size of each group}$$

$$s_W^2 = \frac{\Sigma (n - 1)s^2}{\Sigma (n - 1)}$$

$$\bar{x}_{GM} = \frac{\Sigma x}{N} \qquad N = \Sigma n$$

dfN = k − 1
dfD = N − k

Remember, if the sample means and variances are not given, you can find them with the following formulas:

$$\bar{x} = \frac{\Sigma x}{n} \qquad\qquad s^2 = \frac{\Sigma x^2 - \dfrac{(\Sigma x)^2}{n}}{n - 1}$$

The Scheffè Test

The **Scheffè test** is used if there is a difference among three or more means and the sample sizes are different. The Scheffè test compares two means at a time to see where the differences are. The test is one tailed right, so if the test value is larger than the critical value, there is a difference between the 2 means tested. The critical value for the Scheffè test is called F′.

The critical value for the Scheffè test:

$$F' = (k - 1)(\text{F table value}) \text{ for the desired } \alpha \text{ and}$$

$\text{dfD} = k - 1$ where k is the number of groups
$\text{dfD} = N - k$ where N is the sum of the sample sizes for all groups

---

The test value for the Scheffè test:

$$F_s = \frac{(\bar{x}_i - \bar{x}_j)^2}{s_W^2\left(\frac{1}{n_i} + \frac{1}{n_j}\right)}$$ where $\bar{x}_i$ and $\bar{x}_j$ are the sample means to be compared and n_i and n_j are the sample sizes

$$s_W^2 = \frac{\Sigma(n - 1)s^2}{\Sigma(n - 1)}$$

The Tukey Test

The **Tukey test** is used instead of the Scheffè test if the sample sizes are all the same.

The critical value for the Tukey test:

q = table value for Table N where k = number of groups and df = $v = N - k$ where N is the sum of all the sample sizes

---

The test value for the Tukey test:

$$q = \frac{\bar{x}_i - \bar{x}_j}{\sqrt{\frac{s_W^2}{n}}}$$ where $\bar{x}_i$ and $\bar{x}_j$ are the sample means to be compared

$$s_W^2 = \frac{\Sigma(n - 1)s^2}{\Sigma(n - 1)}$$

Two Way Analysis of Variance

Two Way Analysis of Variance is used to test the effect of 2 independent variables on one dependent variable if the samples are normally distributed, the variances of the populations are equal, and the groups are of equal size.

A and B are the independent variables. A contingency table can be made using as many boxes as are needed.

	B_1	B_2
A_1		
A_2		

There are three sets of hypotheses for a two way analysis of variance.

1. H_0: There is no interaction between A and B on the dependent variable.
    H_1: There is an interaction.

2. H_0: A has no effect on the dependent variable.
    H_1: A has an effect on the dependent variable.

3. H_0: B has no effect on the dependent variable.
    H_1: B has an effect on the dependent variable.

NOTES:

Checking Your Understanding

Complete this section before you do the exercises to make sure you understand the concepts. Write in the book. The answers are in the back of the book. Make a note of any questions that you wish to discuss with the instructor.

I. Discuss the conditions and uses for the F test for 2 variances.

II. Discuss the conditions and uses for ANOVA.

III. Discuss the conditions and uses for the Scheffè test and the Tukey test.

QUESTIONS:

Applying Your Understanding

STUDENT: In the preceding sections, you learned the concepts of the F test and ANOVA and checked your understanding of the concepts. In this section, you will apply your understanding of the concepts. Study each example carefully and then try to work the following exercise. If you have any problems, see your instructor. The answers are in the back of the book.

Example 1--Two variances.

A sample of 25 women had a variance in IQ scores of 162. A sample of 18 men had a variance of 72. Do the women have a larger variance in IQ scores at the .10 level? Assume both populations are normally distributed.

Solution

1. State the hypotheses.
   H_0: $\sigma_1^2 \leq \sigma_2^2$
   H_1: $\sigma_1^2 > \sigma_2^2$

2. Find the critical value and draw a graph.
   In Table H, find the table for .10.
   Label the larger variance s_1^2.
   $s_1^2 = 162$ $\qquad$ $s_2^2 = 72$

   $n_1 = 25$ $\qquad$ $n_2 = 18$

   $dfN = n_1 - 1 = 25 - 1 = 24$ $\qquad$ $dfD = n_2 - 1 = 18 - 1 = 17$

   Table value = 1.84

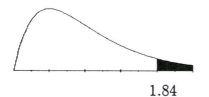

   1.84

3. Compute the test value.

   $$F = \frac{s_1^2}{s_2^2} = \frac{162}{72} = 2.25$$

4. Reject the null.

5. The variances in IQ's are larger for women than for men.

Exercise 1

The variances for the life expectancy of batteries for two different brands are given below. Is the variance for brand A greater than for brand B at the .01 level?

Brand A	Brand B
$s^2 = 129$	$s^2 = 36$
$n = 15$	$n = 12$

Example 2--Two variances.

The standard deviation for daily sales are given for 2 stores. Is there a difference in the variances at the .05 level?

Store X	Store Y
$s = 29$	$s = 36$
$n = 15$	$n = 12$

Solution

1. State the hypotheses.
$$H_0: \sigma_1^2 = \sigma_2^2$$
$$H_1: \sigma_1^2 \neq \sigma_2^2$$

2. Find the critical value and draw the graph.

Two tailed, so find $\frac{\alpha}{2} = \frac{.05}{2} = .025$

Label the larger standard deviation s_1. Square the standard deviations to get the variances.

$$s_1^2 = 36^2 = 1296 \qquad s_2^2 = 29^2 = 841$$
$$n_1 = 12 \qquad\qquad n_2 = 15$$

$$\text{dfN} = n_1 - 1 = 12 - 1 = 11$$
$$\text{dfD} = n_2 - 1 = 15 - 1 = 14$$

Look at Table H and find the table for .025. Since dfN = 11 is not on the table, take the next smaller number, dfN = 10, dfD = 14, table value = 3.15.

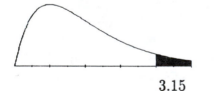

3.15

This is the graph for all F tests of variances, even two tailed.

3. Find the test value.

$$F = \frac{s_1^2}{s_2^2} = \frac{1296}{841} = 1.541$$

4. Accept the null.

5. The variances are equal.

Exercise 2

Cholesterol level was checked for two groups, one over 50 years old, and the other under 50. At the .01 level, is there a difference in the variances?

Over 50	Under 50
$s = 9$	$s = 17$
$n = 30$	$n = 24$

Example 3--ANOVA.

Samples of three brands of batteries have the following times of use before they fail. At the .05 level, is there a difference in the mean lengths of time for the three brands?

Brand A	Brand B	Brand C
30	42	30
14	28	14
22	20	20
18	35	16
26	49	15
25	28	
	36	
	24	

Solution

You must find each mean and variance. Find the sum of each column and the sum of each column squared.

$$\bar{x} = \frac{\Sigma x}{n} \qquad\qquad s^2 = \frac{\Sigma x^2 - \frac{(\Sigma x)^2}{n}}{n-1}$$

Brand A		Brand B		Brand C	
x	x^2	x	x^2	x	x^2
30	900	42	1764	30	900
14	196	28	784	14	196
22	484	20	400	20	400
18	324	35	1225	16	256
26	676	49	2401	15	225
25	625	28	784	95	1977
135	3205	36	1296		
		24	576		
		262	9230		

$$\bar{x}_1 = \frac{135}{6} = 22.50 \qquad \bar{x}_2 = \frac{262}{8} = 32.75 \qquad \bar{x}_3 = \frac{95}{5} = 19.00$$

$$s_1^2 = \frac{3205 - \frac{135^2}{6}}{5} = 33.5 \quad s_2^2 = \frac{9230 - \frac{262^2}{8}}{7} = 92.8 \quad s_3^2 = \frac{1977 - \frac{95^2}{5}}{4} = 43$$

$$n_1 = 6 \qquad\qquad n_2 = 8 \qquad\qquad n_3 = 5$$

1. State the hypotheses.
   H_0: $\mu_1 = \mu_2 = \mu_3$
   H_1 : At least one mean is different.
2. Find the critical value and draw a graph.
   k = number of groups = 3
   $N = n_1 + n_2 + n_3 = 6 + 8 + 5 = 19$
   $\alpha = .05$

   dfN $= K - 1 = 3 - 1 = 2$
   dfD $= N - k = 19 - 3 = 16$
   table value $= 3.63$

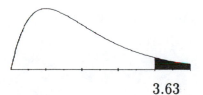

3.63

3. Compute the test value.

$$F = \frac{s_B^2}{s_W^2} \qquad \text{where}$$

$$s_B^2 = \frac{\Sigma n(\bar{x} - \bar{x}_{GM})}{k - 1} \qquad \bar{x} = \text{mean of each group} \qquad n = \text{sample size of each group}$$

$$s_W^2 = \frac{\Sigma(n - 1)s^2}{\Sigma(n - 1)}$$

$$\bar{x}_{GM} = \frac{\Sigma x}{N} \qquad N = \Sigma n$$

$$\bar{x}_{GM} = \frac{135 + 262 + 95}{19} = 25.89$$

$$s_B^2 = \frac{6(22.5 - 25.89)^2 + 8(32.75 - 25.89)^2 + 5(19 - 25.89)^2}{2} = 341.39$$

$$s_W^2 = \frac{5(33.5) + 7(92.8) + 4(43)}{5 + 7 + 4} = 61.82$$

$$F = \frac{341.39}{61.82} = 5.52$$

4. Reject the null.

5. At least one mean is different from the others.

Exercise 3

A department store has three chain stores in three different locations in one town. The following numbers represent the dollars spent per customer per visit. At the .01 level, is there a difference among the means?

Store 1	Store 2	Store 3
15	18	10
12	29	36
25	36	28
36	41	19
40	15	25
	26	13
		12

Example 4--Scheffè test.

Use the data in Example 3 to find between which brands there is a difference at the .05 level.

Solution

The Scheffè test can be used since there is a difference among the means.

$$F_s = \frac{(\bar{x}_i - \bar{x}_j)^2}{s_W^2\left(\frac{1}{n_i} + \frac{1}{n_j}\right)} \qquad \text{where } \bar{x}_i \text{ and } \bar{x}_j \text{ are the sample means to be compared and } n_i \text{ and } n_j \text{ are the sample sizes}$$

dfN $= k - 1 = 3 - 1 = 2$
dfD $= N - k = 19 - 3 = 16$
$\alpha = .05$
table value $= 3.63$

Critical value $= (k - 1)(\text{table value}) = 2(3.63) = 7.26$

There is a difference between any pair of means if the test value is greater than 7.26.

From Example 3:

$\bar{x}_1 = 22.5$ $\qquad\qquad \bar{x}_2 = 32.75$ $\qquad\qquad \bar{x}_3 = 19.0$
$n_1 = 6$ $\qquad\qquad\quad n_2 = 8$ $\qquad\qquad\quad n_3 = 5$

A. $\bar{x}_1$ versus $\bar{x}_2$

$$F = \frac{(22.5 - 32.75)^2}{61.82\left(\frac{1}{6} + \frac{1}{8}\right)} = 5.83$$

B. $\bar{x}_2$ versus $\bar{x}_3$

$$F = \frac{(32.75 - 19)^2}{61.82\left(\frac{1}{8} + \frac{1}{5}\right)} = 9.41$$

C. $\bar{x}_1$ versus $\bar{x}_3$

$$F = \frac{(22.5 - 19)^2}{61.82\left(\frac{1}{6} + \frac{1}{5}\right)} = .54$$

There is a difference between Brand B and Brand C.

QUESTIONS:

QUESTIONS:

Practice Test

1. A sample of 18 females had a variance of 212 on a math test. A sample of 24 males had a variance of 85. Is there a difference in the variances at the .05 level?

2. Three different computer programs were used to solve a statistics problem. At the .01 level, test the claim that the mean times for all three programs to solve the problem is the same. The following table represents the number of minutes per trial for each program.

A	B	C
7	9	2
3	12	3
2	8	4
5		5
4		3
8		2

CHAPTER 14
NONPARAMETRIC STATISTICS
Understanding Nonparametric Statistics

Introduction

Nonparametric statistics are tests that are used when the populations are not normally distributed or to test hypotheses involving medians or ranks. The advantages and disadvantages of nonparametric tests are listed below.

Advantages of Nonparametric Tests:
1. Can be used when the population is not normally distributed.
2. Can be used on nominal or ordinal data.
3. Can be used to test other hypotheses besides population parameters.
4. Are usually easier to compute and understand.

Disadvantages of Nonparametric Tests:
1. Are less sensitive. Larger differences are needed to reject the null.
2. Use less information.
3. Are less efficient so larger sample sizes are required.

Ranking

Many times parameters must be ranked to perform a nonparametric test. To rank a set of numbers, put the numbers in order and rank the smallest number 1, the second 2, and so on. If two numbers are the same in the fourth place, for instance, the rank would be 4.5 for both numbers and the next number would be ranked 6.

Sign Tests

A **single sample sign test** is used to test the value of a median. Numbers in a sample are assigned a $+$ if they are above the assumed median, a $-$ if they are below the assumed median, and a 0 if they are equal to the assumed median. The test value is the number of $+$ or $-$ signs, whichever is smaller. If n is less than or equal to 25, Table J is used to find the critical value for the desired α for one or two tailed test and for n = the total number of $+$ and $-$ signs, zeros are not counted. Reject the null if the test value is less than or equal to the table value. If the sample size is greater than 25, the **normal approximation** can be used to test a median. The critical value can be found on the bottom row of Table F for the desired α. The hypotheses and their graphs are given below.

$H_0: \geq$ $H_0: \leq$ $H_0: =$
$H_1 <$ $H_1: >$ $H_1: \neq$

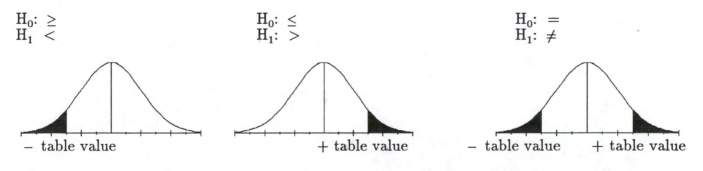

  − table value   + table value   − table value   + table value

The test value for testing the median when $n > 25$

$$z = \frac{(x + 0.5) - \frac{n}{2}}{\frac{\sqrt{n}}{2}}$$

x = number of + or − signs, whichever is smaller
n = sample size

A **paired sample sign test** can be used to compare two dependent samples if the populations are not normally distributed. The + and − signs are obtained by subtracting each value in the second group from the corresponding value in the first group and putting +, −, or 0 for each difference. The critical value is obtained from Table J for the desired α and n = number of + and − signs (do not count the zeros). The test value is the number of + or − signs, whichever is smaller. Reject the null if the test value is smaller than or equal to the critical value. The normal approximation can be used if $n > 25$.

The **Wilcoxon rank sum test** combines the values from both samples and ranks them. The null hypothesis is that there is no difference in the population. The alternate is that there is a difference. Both samples must be larger than 10. The critical value is the z for Table F for a two tailed test. If the test value is below the −table value or above the + table value, reject the null.

The test value for Wilcoxon rank sum test

$$z = \frac{R - \mu_R}{\sigma_R}$$

$$\mu_R = \frac{n_1(n_1 + n_2 + 1)}{2}$$

$$\sigma_R = \sqrt{\frac{n_1 n_2(n_1 + n_2 + 1)}{12}}$$

n_1 = smaller of the sample sizes (If the sample sizes are equal, either can be called n_1.)
n_2 = larger sample size
R = sum of ranks from the smaller sample size

The **Wilcoxon signed rank test** is used for dependent samples and ranks the differences between the first set of numbers and the second set of numbers. The procedures are given below.

Wilcoxon Signed Rank Test
Step 1: State the hypotheses.
Step 2: Find the critical value from Table K for the desired α and either one or two tailed.
        Reject the null if the test value is greater than or equal to the critical value.
Step 3: Find the critical value.
        (a) Make a table.

Before	After	D	|D|	Rank	Signed Rank
x_B	x_A	$x_B - x_A$			

        D = the first value minus the second value for each pair of numbers
        |D| = the absolute value of D
        Rank = the rank for the absolute values of the differences
        Signed rank = the rank with a + or − sign according to the sign in column D
        (b) Find the sum of the positive ranks and the negative ranks separately.
        (c) The test value is the smaller of the absolute values of the sum of the positive or
            the sum of the negative ranks.
Step 4: Make the decision.
Step 5: Summarize the results.

If $n \geq 31$, use Table F to find the z value for the critical value for the Wilcoxon signed rank test and the test value is:

$$z = \frac{w_s - \frac{n(n+1)}{4}}{\sqrt{\frac{n(n+1)(n+2)}{24}}}$$

n = number of pairs where the difference is not 0
w_s = smaller sum of the absolute value of the signed ranks

The **Kruskal-Wallis test** is used to compare means of three or more populations if the populations are not normally distributed or if the variances are not equal. Each sample must have five or more values. The null hypothesis is that there is no difference between the populations. The alternate is that at least one of the populations is different. This is a one tail test, df = $k - 1$ where k is the number of groups. Table G is used to find the critical value. All the values should be placed together and ranked. The R is the sum of the ranks for each group.

Test value for the Kruskal-Wallis or H test

$$H = \frac{12}{N(N+1)}\left(\frac{R_1^2}{n_1} + \frac{R_2^2}{n_2} + \ldots + \frac{R_k^2}{n_k}\right) - 3(N+1)$$

R = sum of the rank for each group
n = size of each sample
$N = n_1 + n_2 + \cdots + n_k$
k = number of samples

Spearman Rank Correlation Coefficient

The **Spearman rank correlation coefficient** is used to test if there is a significant linear correlation between two sets of numbers if the populations are not normally distributed. The null hypothesis is that there is no correlation and the alternate is that there is a correlation. The Greek letter rho, ρ, is used for the correlation. The null hypothesis is $\rho = 0$. The alternate is $\rho \neq 0$. The critical value is obtained from Table L for the desired α and n = the number of pairs of values. Reject the null if the absolute value of the test value is greater than critical value. Each group is ranked separately and d = the rank from the first group minus the rank from the second group.

To test correlation where the populations are not normally distributed, the test value is:

$$r = 1 - \frac{6\sum d^2}{n(n^2 - 1)}$$

d = rank of the first group minus the second group
n = number of pairs of values

Test for Runs

A **test for runs** is used to see if data was selected at random. The data needs to be labeled with two different letters. Nominal level data can be used for a test of runs. If working with people and you are interested if the males and females were selected at random, M could be used for the males and F for the females. If numbers are used, A can be used for the numbers above the median and B for the numbers below the median. If the data was randomly picked, there will be no discernible pattern to the runs of the letters. A **run** is a succession of like letters. A group of random data should not have too few or too many runs. A A B B B A A B A A B B A has 7 runs. Table M is used to find the critical value for a .05 level test. To read Table M, look for the number of values in the first group along the top and the number of values in the second group along the left hand column. It does not matter which group you call the first, the table will give the same amount. The table will give you two values. Accept the null if the test value is between the two table values. If the number of runs is less than or equal to the smaller value or more than or equal to the larger value, reject the null hypothesis. The null hypothesis is that the data is random. The alternate is that the data is not random.

NOTES:

Checking Your Understanding

Complete this section before you do the exercises to make sure you understand the concepts. Write in the book. The answers are in the back of the book. Make a note of any questions that you wish to discuss with the instructor.

True or False

_____ 1. Nonparametric tests can be used when the data is from a population that is not normally distributed.

_____ 2. Nominal level data may not be used in nonparametric tests.

_____ 3. Nonparametric tests use less information.

_____ 4. Nonparametric tests are more sensitive.

_____ 5. Larger sample sizes are needed for nonparametric tests.

_____ 6. A sign test can be used to test hypotheses about a median.

_____ 7. A paired sample test can be used on data that is independent.

_____ 8. The Wilcoxon rank sum test can be used on data that is independent.

_____ 9. A test for runs tests randomness.

QUESTIONS:

Applying Your Understanding

STUDENT: In the preceding sections, you learned the concepts of nonparametric statistics and checked your understanding. In this section, you will apply your understanding of the concepts. Study each example carefully and then try to work the following exercise. If you have any problems, see your instructor. The answers to the exercises are in the back of the book.

Example 1--Testing medians.

A teacher claims the median grade on a test was 72. A sample of 25 students had the following grades. Check the teacher's claim at the .01 level. Assume the population is not normal.

68	52	76	92	96
72	81	91	61	94
74	72	42	87	72
69	78	85	73	81
52	86	73	78	75

Solution

1. State the hypotheses.
   H_0: median $= 72$
   H_1: median $\neq 72$

2. Find the test value.
   Put a $+$ by each number greater than 72, a $-$ by each number smaller than 72, and 0 by each 72.

68 $-$	52 $-$	76 $+$	92 $+$	96 $+$
72 0	81 $+$	91 $+$	61 $-$	94 $+$
74 $+$	72 0	42 $-$	87 $+$	72 0
69 $-$	78 $+$	85 $+$	73 $+$	81 $+$
52 $-$	86 $+$	73 $+$	78 $+$	75 $+$

   There are 6 $-$ and 16 $+$. The smaller value is 6. The test statistic is 6.

3. Find the critical value.
   On Table J, $\alpha = .01$, two tail test, look up $n = 22$ (number of $+$ and $-$ signs, do not count 0's). Critical value $= 4$. Reject the null if the test value is less than or equal to 4.

4. Accept the null.

5. The median is 72.

Exercise 1

Is the median for the IQ scores 112? Test at the .05 level.

110	108	107	111	112	121
102	126	109	104	109	107
112	107	106	109	110	103

Example 2--Paired sample sign test.

A researcher believes that amphetamines can raise IQ scores. A sample of 16 people were given an IQ test. Two months later, the same people were given another test after taking a small dose of an amphetamine. If these people were from a group whose IQ scores were not normally distributed, did their IQ scores raise at the .01 level?

Before	After
110	111
103	105
112	118
102	104
106	105
109	109
117	119
109	110
110	111
107	108
106	106
104	105
110	112
107	108
109	107
111	112

Solution

1. State the hypotheses.
   H_0: The IQ scores will not be raised.
   H_1: The IQ scores will be raised.

2. Find the test value.
   Subtract the first value from the second score and put a $+$, $-$ or 0 for the difference.

Before	After	Sign of the difference
110	111	$+$
103	105	$+$
112	118	$+$
102	104	$+$
106	105	$-$
109	109	0
117	119	$+$
109	110	$+$
110	111	$+$
107	108	$+$
106	106	0
104	105	$+$
110	112	$+$
107	108	$+$
109	107	$-$
111	112	$+$

   number of $+$ = 12
   number of $-$ = 2
   test value = smaller number of $+$ or $-$ = 2

3. Find the critical value.
   Use Table J, $\alpha = .01$, one tailed, $n = 14$ (number of $+$ and $-$, do not count 0's)
   Critical value = 2
   Reject the null if the test value is less than or equal to 2.

4. Reject the null.
5. IQ scores were raised.

> **Exercise 2**
>
> 16 workers in an office volunteered to have their stress levels tested. On a Monday morning their levels were tested. Their stress levels were tested again on Thursday morning. Did the weekend lower their stress level? (Is the stress level lower on Monday than on Thursday?) Test at the .05 level and assume that stress levels are not normally distributed.
>
> | Monday: | 68 | 86 | 31 | 84 | 72 | 86 | 91 | 47 | 86 | 74 | 81 | 82 | 76 | 68 | 51 | 61 |
> | Tuesday: | 69 | 87 | 35 | 80 | 70 | 88 | 93 | 47 | 88 | 72 | 83 | 88 | 78 | 70 | 57 | 65 |

Example 3--Wilcoxon rank sum test.

The time it took a sample of females and males to complete a math test were compared. Is there a difference in the means at the .05 level? The females and males are from independent samples and the times of completion are not normally distributed.

Females	Males
28	49
36	48
42	44
50	38
48	35
50	36
30	45
29	46
24	29
43	40
38	47
	49
	47

Solution

1. State the hypotheses.
   H_0: The distributions are the same.
   H_1: The distributions are not the same.

2. Find the critical value and draw a graph.
   $\alpha = .05$, two tail test, Table F, critical value $= 1.96$

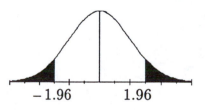

3. Find the test value.
   $$z = \frac{R - \mu_R}{\sigma_R}$$
   $$\mu_R = \frac{n_1(n_1 + n_2 + 1)}{2}$$
   $$\sigma_R = \sqrt{\frac{n_1 n_2(n_1 + n_2 + 1)}{12}}$$

n_1 = smaller of the sample sizes = 11
n_2 = larger sample size = 13
R = sum of ranks from the smaller sample size

List all the numbers in order and rank them. Keep track of which are male (M) and female (F).

		Rank
F	24	1
F	28	2
F	29	3.5
M	29	3.5
F	30	5
M	35	6
M	36	7.5
F	36	7.5
F	38	9.5
M	38	9.5
M	40	11
F	42	12
F	43	13
M	44	14
M	45	15
M	46	16
M	47	17.5
M	47	17.5
M	48	19.5
F	48	19.5
M	49	21.5
M	49	21.5
F	50	23.5
F	50	23.5

There were two number 29's, so they were each ranked 3.5, and the next number is ranked 5.

R = sum of ranks of the group with the smaller sample size = sum of ranks of the female
= 1 + 2 + 3.5 + 5 + 7.5 + 9.5 + 12 + 13 + 19.5 + 23.5 + 23.5 = 120

$$\mu_R = \frac{n_1(n_1 + n_2 + 1)}{2} = \frac{11(11 + 13 + 1)}{2} = \frac{275}{2} = 137.5$$

$$\sigma_R = \sqrt{\frac{n_1 n_2 (n_1 + n_2 + 1)}{12}} = \sqrt{\frac{11(13)(11 + 13 + 1)}{12}} = \sqrt{\frac{3575}{12}} = \sqrt{297.91667} = 17.26$$

$$z = \frac{R - \mu_R}{\sigma_R} = \frac{120 - 137.5}{17.26} = -1.01$$

4. Accept the null.

5. There is no difference.

Exercise 3

A store wishes to see if there is a difference in the amount spent by customers using a credit card and the amount spent by customers paying in cash. Test at the .05 level.

Credit Card	Cash
16	26
23	38
46	42
38	18
23	17
29	28
36	47
48	52
28	36
21	42
25	47
	51

Example 4--The Kruskal-Wallis test.

Reaction times were tested for three different age groups. Is there a difference in reaction times among the different age groups at the .05 level? Assume the populations are not normally distributed.

Ages:	10-19	20-29	30-39
	12	18	14
	19	13	23
	17	21	27
	21	35	31
	25	28	29
	26	30	32

Solution

1. State the hypotheses.
   H_0: The populations are the same.
   H_1: The populations are not the same.

2. Find the critical value.
   one tail right, $\alpha = .05$, df $= k - 1 =$ number of groups $- 1 = 3 - 1 = 2$, from Table G critical value $= 5.991$ Reject the null if the test value is larger than 5.991

3. Find the test value.

$$H = \frac{12}{N(N+1)}\left(\frac{R_1^2}{n_1} + \frac{R_2^2}{n_2} + \ldots + \frac{R_k^2}{n_k}\right) - 3(N+1)$$

R = sum of the rank for each group
n = size of each sample
$N = n_1 + n_2 + \cdots + n_k$
k = number of samples

Put all the numbers in order and rank them. Keep track of which group they came from. (10-19 Group I, 20-29 Group II, 30-39 Group III) Sum the ranks for each group.

Group	Number	Rank
I	12	1
II	13	2
III	14	3
I	17	4
II	18	5
I	19	6
II	21	7.5
I	21	7.5
III	23	9
I	25	10
I	26	11
III	27	12
II	28	13
III	29	14
II	30	15
III	31	16
III	32	17
II	35	18

Group I	Rank	Group II	Rank	Group III	Rank
12	1	16	5	14	3
19	6	13	2	23	9
17	4	21	7.5	27	12
21	7.5	35	18	31	16
25	10	28	13	29	14
26	11	30	15	32	17
	$R_1 = 39.5$		$R_2 = 60.5$		$R_3 = 71$
	$n_1 = 6$		$n_2 = 6$		$n_3 = 6$

$$N = 6 + 6 + 6 = 18$$

$$H = \frac{12}{(18)(19)}\left(\frac{39.5^2}{6} + \frac{60.5^2}{6} + \frac{71^2}{6}\right) - 3(19) = 3.01$$

4. Accept the null.

5. The populations are the same. Age does not make a difference.

Exercise 4

Three different shifts work at a factory. If the populations are not normally distributed, is there a difference among the amounts produced by the different shifts? Test at the .01 level.

Morning Shift	Evening Shift	Night Shift
231	222	220
216	225	205
235	224	201
241	230	212
236	227	216
239	221	214

Example 5--Spearman rank correlation coefficient.

Six teachers were ranked by their students and by their supervisor. At the .05 level, is there a correlation between the students' and the supervisor's rankings? Assume the population is not normally distributed.

Teacher	Students' ranks	Supervisor's rank
A	5	3
B	2	4
C	1	1
D	4	5
E	3	2
F	6	6

Solution

1. State the hypotheses.
   H_0: $\rho = 0$ (no correlation)
   H_1: $\rho \neq 0$ (there is a correlation)

2. Find the critical value.
   Table L, $a = .05$, $n = 6$, Table value = .886
   Reject the null if the absolute value of the test value is larger than .866.

3. Find the test value.

   $$r_s = 1 - \frac{6 \sum d^2}{n(n^2 - 1)} \qquad d = \text{difference in ranks} \qquad n = \text{number of pairs of numbers}$$

Rank	Rank	d	d^2
5	3	2	4
2	4	−2	4
1	1	0	0
4	5	−1	1
3	2	1	1
6	6	0	0
			10

   $$r_s = 1 - \frac{6(10)}{6(36 - 1)} = .714$$

4. Accept the null.

5. There is no correlation.

Exercise 5

Seven different contestants were rated by 2 judges on a scale of 1 to 10. If the populations are not normally distributed, is there a correlation at the .05 level?

Contestant	Judge 1	Judge 2
A	9	8
B	7	5
C	10	9
D	9	8
E	6	5
F	8	8
G	9	9

Example 6--Test for runs.

A teacher listed test grades in the order that the papers were turned in. At the .05 level, was the order the papers were turned in random?

71   87   86   70   69   72   74   86   62   67   68   69   96   70   63   74   92   63   75

Solution

1. State the hypotheses.
   H_0: The grades were in random order.
   H_1: The grades were not in random order.

2. Find the test value.
   Find the median and put A for each number above the median, B for each number below the median, and nothing for each number that is the same as the median. To find the median, the numbers must be put in order the find the middle number.

   62   63   63   67   69   70   70   71   72   74   74   75   86   86   87   92   96
   median = 71
   71   87   86   70   69   72   74   86   62   67   68   69   96   70   63   74   92   63   75
         A   A   B   B   A   A   A   B   B   B   B   A   B   B   A   A   B   A

   There are 9 runs.

3. Find the critical value.
   Use Table M, n_1 = number of A's = 9, n_2 = number of B's = 9
   Table values = 5 and 15
   Accept the null if the test value is between the table values.

4. Accept the null.

5. The papers were in random order by grade.

Exercise 6

A theater wanted to know if the people waiting in line to by tickets were in random order according to sex. Test at the .01 level.

M F M F M M M F M F M F F F M F M M F F M F M F M F M F M F M F F F M

QUESTIONS:

Practice Test

1. A record store employee claims that the median age of the customers is 17. Test at $\alpha = .01$.

   15 16 14 13 12 11 18 19 20 21 25 27 16 15 14 12

2. A group of dieters started an exercise program. Is there a difference in the weight loss without and with exercise at the .05 level. Use the paired sample test.

   Average weight loss per week

Dieter	No exercise	With exercise
1	2	3
2	3	3.5
3	1	1.5
4	2.5	3
5	4	6
6	2	4
7	4	3
8	5	6.5
9	5	5
10	4	6.5

3. A group of men and women smokers were asked at what age they started smoking. Using the Wilcoxon rank sum test at the .01 level, determine if there is a significant difference in the starting ages based on gender.

Females	Males
14	14
15	12
22	15
18	16
17	16
22	13
19	12
20	18
21	17
13	19

4. Students were asked to rate a teacher on a scale of 1 to 20. The students were grouped by the grade they made in the class. Is there a difference in the ranks given by students with different grades? Test at the .01 level. Assume the populations are not normally distributed, so use the Kruskal-Wallis test.

A or B	C	D or F
16	19	12
17	18	14
18	16	15
15	14	10
	19	12
	17	

5. Six different soft drinks were ranked by children and adults. Use the Spearman rank correlation coefficient to test if there is a relationship between how adults and children rank soft drinks.

Brand	Children	Adults
A	1	2
B	2	1
C	5	5
D	4	4
E	3	3
F	6	6

6. Are these answers for a true-false test in random order?

T T F T F F T T T F F T F F T T F F T T

QUESTIONS:

CHAPTER 15
SAMPLING AND SIMULATION
Understanding Sampling and Simulation

Sampling

Sampling techniques are used when part of a population is to be surveyed. If it takes too long or is too expensive to interview or survey the whole population, a sample is used. If a sample is chosen correctly so it represents the population, it is called **unbiased**. If the sample is chosen so the whole population is not fairly represented, it is **biased**.

Random sampling is used to see that all the possible elements of the population have an equal chance of being selected. One method to obtain a random sample is to number all the elements in a population, mix the numbers thoroughly and draw out how many elements you want to sample and interview just the ones whose numbers were chosen. Another way to choose the numbers without mixing and drawing is to use a random number chart. Table D is a random chart. Close your eyes and point the chart to see what number to start with. If you want to sample 10 elements, write down 10 numbers starting with the one you picked and sample those 10 elements. Notice the table has 5 digits in each number. You read as many of the digits as you need. If you only need 2 digits, take the last 2 digits from the numbers you picked. If you need more than 5 digits, read 2 columns and take as many digits as you need.

Systematic sampling numbers the population and then selects numbers at a regular interval. Depending on how many you want to sample, you might take every third or every tenth number. To use systematic sampling, the population must be randomly mixed when you number them.

Stratified sampling divides the population into subgroups that have characteristics that might be important to the study. A population might be divided according to sex, age, or income level, if these are important to the study. Then a proportional number from each group is chosen to ensure a fair representation in the sample.

Cluster sampling samples an already existing group. One class might be used as a sample for a whole school. Cluster sample is often easier and cheaper than other methods, but sometimes the clusters do not represent the whole group.

Sequence sampling test successive units. Businesses use sequence sampling to make sure production items all meet certain standards. If you want to make sure that items produced at the end to the day are as good at items produced at the first of the day, all the items for one day might be tested and compared to the time of day produced.

Double sampling first uses a questionnaire to pick a group with the characteristics to be studied and then studies a sample from that group. **Multistage sampling** uses a combination of sampling techniques. For example, a random sample might be picked from a cluster.

Simulation

Simulation techniques use probabilities and random numbers to simulate real life problems. the **Monte Carlo method** of simulation finds the probability for each of the possible outcomes of an experiment and uses random numbers to simulate real-life situations.

Steps for Simulation

1. List all possible outcomes of an experiment.
2. Determine the probabilities of each outcome.

3. Set up a correspondence between the outcomes of the experiments and the random numbers.
4. Select random numbers from a table and conduct the experiment.
5. Repeat the experiment and tally the outcomes.
6. Compute any statistics and state the conclusions.

NOTES:

Checking Your Understanding

Complete this section before you do the exercises to make sure you understand the concepts. Write in the book. The answers are in the back of the book. Make a note of any questions you wish to discuss with the instructor.

I. In a bakery, a machine is filled with raw cookie dough and 50 cookies are pressed out. Then the machine is filled again and presses out 50 more cookies. 10,000 cookies can be pressed out in one day. The weights of the cookies need to be tested to insure that the machine is working properly. Explain how to use each of the following sample techniques to test the weights of the cookies and discuss why each method may or may not be appropriate.

    a. random sampling

    b. systematic sampling

    c. stratified sampling

    d. cluster sampling

    e. sequence sampling

II. Select the correct answer and write the appropriate letter in the space provided.

_____ 1. A sample that does not represent a population correctly is called
    a. random.
    b. clustered.
    c. biased.
    d. unbiased.

_____ 2. Techniques that use random numbers and probabilities to represent real life situations are called
   a. stratified.
   b. sampling.
   c. simulation.
   d. systematic.

_____ 3. Sampling that subdivides the population into subgroups is called
   a. clusters.
   b. random.
   c. stratified.
   d. sequence.

_____ 4. Using preexisting groups in sampling is called
   a. stratified.
   b. random.
   c. systematic.
   d. clusters.

_____ 5. If you sample every third item, you would be using
   a. random sampling.
   b. sequence sampling.
   c. cluster sampling.
   d. systematic sampling.

_____ 6. If you sample 50 items in a row you would be using
   a. random sampling.
   b. sequence sampling.
   c. cluster sampling.
   d. double sampling.

_____ 7. Number all the items and then drawing numbers from a hat to determine which ones to test would be called
   a. random sampling.
   b. sequence sampling.
   c. cluster sampling.
   d. double sampling.

QUESTIONS:

Applying Your Understanding

STUDENT: In the preceding sections, you learned the concepts of sampling and simulation, and checked your understanding of the concepts. Study each example carefully and then try to work the following exercise. If you have any problems, see your instructor. The answers to the exercises are in the back of the book.

Use this chart of cities with population of 50,000 to 80,000 with the population and number of new dwellings constructed in one year for examples 1-4.

City Number	Population	New Dwellings	City Number	Population	New Dwellings
1.	64.480	103	18.	61,657	981
2.	71,508	243	19.	50,576	406
3.	52,523	289	20.	62,860	170
4.	70,174	135	21.	74,549	547
5.	66,269	389	22.	54,263	132
6.	55,725	977	23.	76,568	322
7.	57,951	475	24.	50,676	696
8.	54,661	321	25.	63,232	282
9.	57,704	139	26.	68,071	383
10.	63,774	192	27.	51,910	287
11.	66,113	157	28.	78,899	999
12.	58,479	307	29.	73,726	645
13.	53,112	802	30.	57,702	103
14.	73,681	919	31.	63,685	584
15.	65,679	683	32.	52,576	858
16.	66,731	763	33.	65,198	636
17.	58,891	130	34.	50,221	122

Example 1--Random sampling.

Use random sampling to find the mean population and mean number of new dwellings for a sample of 10 cities.

Solution

Use a random number table and put your finger or pen down in any spot. Since there are only 34 numbers in our population, use only the last two digits of each number. Any number more than 34 is disregarded and drop any duplicates. A random selction might give you:
08   34   27   18   26   21   33   16   31   01
Use these cities and find the mean population and mean number of new dwellings.

City	Population	New Dwellings
8.	54,661	321
34.	50,221	122
27.	51,910	287
18.	61,657	981
26.	68,071	383
21.	74,549	547
33.	65,198	636
16.	66,731	763
31.	63,685	584
1.	64,480	103
	621,163	4727

For population: $\bar{x} = \dfrac{621,163}{10} = 62,116.3$

For new dwellings: $\bar{x} = \dfrac{4727}{10} = 472.7$

Exercise 1

Use random sampling to pick 10 people and find the mean age and mean number of soft
drinks drank per week.   Soft drinks   Soft drinks

Number	Age	Drank per week	Number	Age	Drank per week
1.	33	5	26.	19	7
2.	70	1	27.	18	9
3.	71	2	28.	20	15
4.	28	14	29.	18	10
5.	35	4	30.	22	20
6.	13	5	31.	21	3
7.	48	1	32.	29	10
8.	32	14	33.	61	3
9.	50	5	34.	57	4
10.	43	20	35.	65	4
11.	40	5	36.	53	5
12.	49	2	37.	58	7
13.	23	16	38.	42	8
14.	55	3	39.	60	14
15.	15	23	40.	50	7
16.	36	12	41.	73	1
17.	19	35	42.	37	9
18.	37	1	43.	22	20
19.	46	2	44.	47	4
20.	27	2	45.	24	14
21.	23	25	46.	43	4
22.	23	1	47.	16	25
23.	17	21	48.	21	20
24.	21	1	49.	20	25
25.	19	9	50.	19	15

Example 2--Systematic sampling.

Select 10 cities by the systematic method and find the mean population and the mean
number of new dwellings.

Solution

Since there are 34 cities and you want to pick 10, pick every third number ($\frac{34}{10} = 3.4$, 3rd or

4th number). You can pick 1, 2, or 3 for the first number. If you pick 2 as the first
number, use 2, 5, 8, 11, and so on.

City	Population	New Dwellings
2.	71,508	243
5.	66,269	389
8.	54,661	321
11.	66,113	157
14.	73,681	919
17.	58,891	130
20.	62,860	170
23.	76,568	322
26.	68,071	383
29.	73,726	645
	672,348	3679

For population: $\bar{x} = \dfrac{672,348}{10} = 67,234.8$

For new dwellings: $\bar{x} = \dfrac{3679}{10} = 367.9$

Exercise 2

Use systematic sampling to pick 10 people and find the mean age and number of soft drinks drank per week for the data in Exercise 1.

Example 3--Stratified sampling.

Use the population to divide the cities into groups. Decide how many from each group to take a sample of 10 and find the mean number of new dwellings.

Solution

50-59,999	New Dwellings	60-69,999	New Dwellings	70-79,999	New Dwellings
3.	289	1.	103	2.	243
6.	977	5.	389	4.	135
7.	475	10.	192	14.	919
8.	321	11.	157	21.	547
9.	139	15.	683	23.	322
12.	307	16.	763	28.	999
13.	802	18.	981	29.	645
17.	130	20.	170		
19.	406	25.	282		
22.	132	26.	383		
24.	696	31.	584		
27.	287	33.	636		
30.	103				
32.	858				
34.	122				

Find the percent that are in each group. Group 1 has 15 cities or $\frac{15}{34} = .44 = 44\%$

Group 2 has 12 cities or $\frac{12}{34} = 35\%$ Group 3 has 7 cities or $\frac{7}{34} = 21\%$

The sample should have the same percentage from each group as the population does. You want 10 in the sample.
Group 1: 35% of 10 = .35(10) = 3.5   Take 3 or 4.
Group 2: 44% of 10 = .44(10) = 4.4   Take 4.
Group 3: 21% of 10 = .21(10) = 2.1   Take 2.
You need to take 4 from Group 1 to make 10.

To randomly pick 4 from group 1, use a random number chart or put all 15 numbers in a hat and draw out 4. Similarly pick 4 from group 2 and 2 from group 3.

Cities	New Dwellings
9.	139
32.	858
13.	802
6.	977
10.	192
15.	683
20.	170
18.	981
21.	547
2.	243
	5592

$$\bar{x} = \frac{5592}{10} = 559.2$$

Exercise 3

Divide the data in Exercise 1 into groups by age. Use 0 - 19 years, 20 - 39 years, 40 - 59 years, and 60 - 79 years. Decide how many from each group and take a stratified sample to the find the mean number of soft drinks.

Example 4--Sequence sampling.

Select 10 cities in sequence and find the mean population and mean number of new dwellings.

Solution

Select any number to start with and select 10 cities in order. A random number table can be used to pick the starting number. If the number you pick is 28, then take cities 28, 29, 30 and so on. If you run out of cities before you get 10 for your sample, go back to number 1.

Cities	Population	New Dwellings
28.	78,899	999
29.	73,726	645
30.	57,702	103
31.	63,685	584
32.	52,576	858
33.	65,198	636
34.	50,221	122
1.	64,480	103
2.	71,508	243
3.	52,523	289
	630,518	4582

For population: $\bar{x} = \dfrac{630,518}{10} = 63,051.8$ For new dwellings: $\bar{x} = \dfrac{4582}{10} = 458.2$

Exercise 4

Pick 10 people in a sequence in Exercise 1 and find the mean age and mean number of soft drinks drank.

Practice Test

Use this chart that shows the number of hours worked per week and the number of credit hours enrolled in for 50 students with part time jobs.

Student	Work	Credit Hours	Student	Work	Credit Hours
1.	20	14	26.	38	9
2.	28	13	27.	35	12
3.	13	16	28.	10	18
4.	20	16	29.	12	21
5.	20	18	30.	10	18
6.	20	15	31.	20	12
7.	15	6	32.	20	17
8.	38	13	33.	20	12
9.	24	13	34.	40	12
10.	20	17	35.	20	18
11.	22	13	36.	28	13
12.	12	9	37.	12	9
13.	28	13	38.	24	18
14.	3	12	39.	24	18
15.	8	6	40.	6	18
16.	38	13	41.	10	17
17.	20	18	42.	40	13
18.	20	15	43.	4	13
19.	20	21	44.	18	16
20.	38	13	45.	25	13
21.	36	12	46.	20	17
22.	10	9	47.	32	17
23.	28	13	48.	12	18
24.	25	16	49.	18	12
25.	10	13	50.	15	15

1. Use random sampling to pick 10 students and find the mean number of hours worked and the mean number of credit hours taken.

2. Use systematic sampling to find the mean hours worked and the mean credit hours taken for a sample of 10 students.

3. Start with any student and use a group of 10 students to do a sequence sample to find the mean number of hours worked and the mean number of credit hours.

4. Use the hours worked to divide the students into 4 groups. (Use 1 - 10 hours, 11 - 20 hours, 21 - 30 hours, and 31 - 40 hours.) Decide how many to take from each group to take a stratified sample of 10 and find the mean number of credit hours taken.

CHAPTER 16
QUALITY CONTROL
Understanding Quality Control

Introduction

Quality control charts are used to check the reliability of a process. The measure to be studied can be means, ranges, or proportions. Several small samples are taken to test the process. The average of the measure from all of the samples is called the **grand mean** and is plotted as a horizontal line on a chart. The **upper control limit, UCL,** and **lower control limit, LCL,** are plotted as an upper horizontal and lower horizontal line. The measure from each of the samples is plotted as a point on the chart and the points are connected with a line. If all the samples are within the limits, the process is **in control.** If any of the measures is outside of the limits, the process is **out of control** and needs to be fixed. Two **variable charts** are the $\bar{x}$ chart and the R chart. The $\bar{x}$ chart plots means and the R chart plots ranges.

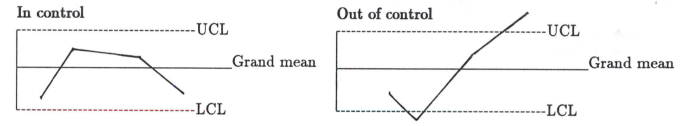

To make a $\bar{x}$ chart

1. Find the mean and range for each sample. The samples must all be the same size, n.

2. Find the grand mean, $\bar{x}_{GM}$, and the mean of the ranges, $\overline{R}$.

$$\bar{x}_{GM} = \frac{\sum \bar{x}}{k} \qquad k = \text{number of samples} \qquad \overline{R} = \frac{\sum R}{k}$$

3. Compute the upper and lower control limits.

$$\text{UCL}_{\bar{x}} = \bar{x}_{GM} + A_2 \overline{R} \qquad A_2 \text{ is obtained from Table O for } n = \text{number in each sample}$$

$$\text{LCL}_{\bar{x}} = \bar{x}_{GM} - A_2 \overline{R}$$

4. Draw a chart using the value for $\bar{x}_{GM}$ for the center line and the values for the $\text{UCL}_{\bar{x}}$ and $\text{LCL}_{\bar{x}}$ as the upper and lower lines.

5. Plot the means for each sample. Draw a line connecting the points.

6. If any mean is outside of the limits, the process is out of control. If all of the means are inside the limits, the process is in control.

To make a $\overline{R}$ chart

1. Find the range for each sample. The samples must all be the same size, n.

2. Find the mean of the ranges, $\overline{R}$. $\qquad \overline{R} = \frac{\sum R}{k} \qquad k = \text{number of samples}$

3. Compute the upper and lower control limits.

$$\text{UCL}_{\text{R}} = \text{D}_4\,\overline{\text{R}}$$ D_4 is obtained from Table O for n = number in each sample

$$\text{LCL}_{\text{R}} = \text{D}_3\,\overline{\text{R}}$$ D_3 is obtained from Table O for n = number in each sample.

4. Draw a chart using the value for $\overline{\text{R}}$ for the center line and the values for the UCL_{R} and LCL_{R} as the upper and lower lines.

5. Plot the ranges for each sample. Draw a line connecting the points.

6. If any range is outside of the limits, the process is out of control. If all of the ranges are inside the limits, the process is in control.

Attribute charts are used if the items to be studied can be classified as acceptable or defective. The $\overline{\text{p}}$ **chart** plots the proportion of defectives in each sample. The $\overline{\text{c}}$ **chart** plots the number of defects for items that are known to be defective.

To make a $\overline{\text{p}}$ chart

1. Find the proportion of defective parts for each sample. The samples must all be the same size, n.

2. Find the mean of the proportions, $\overline{\text{p}}$. $$\overline{\text{p}} = \frac{\sum \text{p}}{k}$$ k = number of samples

3. Compute the upper and lower control limits.

$$\text{UCL}_{\text{p}} = \overline{\text{p}} + 3\sqrt{\frac{\overline{\text{p}}(1 - \overline{\text{p}})}{n}}$$

$$\text{LCL}_{\text{p}} = \overline{\text{p}} - 3\sqrt{\frac{\overline{\text{p}}(1 - \overline{\text{p}})}{n}}$$

4. Draw a chart using the value for $\overline{\text{p}}$ for the center line and the values for the UCL_{p} and LCL_{p} as the upper and lower lines.

5. Plot the proportions for each sample. Draw a line connecting the points.

6. If any proportion is outside of the limits, the process is out of control. If all of the proportions are inside the limits, the process is in control.

To make a $\overline{\text{c}}$ chart

1. Find the average number of defective parts for all the items in the sample.

$$\overline{\text{c}} = \frac{\sum \text{c}}{n}$$ n = number of parts

2. Compute the upper and lower control limits. $\text{UCL}_{\text{c}} = \overline{\text{c}} + 3\sqrt{\overline{\text{c}}}$ $\text{LCL}_{\text{c}} = \overline{\text{c}} - 3\sqrt{\overline{\text{c}}}$

3. Draw a chart using the value for $\overline{\text{c}}$ for the center line and the values for the UCL_{c} and LCL_{c} as the upper and lower lines.

4. Plot the number of defective parts for the items. Draw a line connecting the points.
5. If any number is outside of the limits, the process is out of control. If all of the numbers are inside the limits, the process is in control.

Checking Your Understanding

Complete this section before you do the exercises to make sure you understand the concepts. Write in the book. The answers are in the back of the book. Make a note of any questions that you wish to discuss with your instructor.

True or false.

_____ 1. The $\bar{x}$ chart studies means.

_____ 2. The $\bar{c}$ chart compares proportions.

_____ 3. The $\bar{p}$ chart is an attribute chart.

_____ 4. The R chart is an attribute chart.

QUESTIONS:

QUESTIONS:

Applying Your Understanding

STUDENT: In the preceding sections, you learned the concepts of quality control and checked your understanding. In this section, you will apply your understanding of the concepts. Study each example carefully and then try to work the following exercise. If you have any problems, see your instructor. The answers to the exercises are in the back of the book.

Example 1--$\bar{x}$ chart.

The average weight of cookies is to be studied to make sure the machine is operating properly. Six samples were taken during one day. Make a $\bar{x}$ to see if the the average weights are in control.

Weights in grams

Sample 1: 45 48 50 51
Sample 2: 41 42 43 44
Sample 3: 46 42 48 42
Sample 4: 48 46 42 44
Sample 5: 44 43 44 47

Solution

1. Find the mean and range for each sample.

Sample 1: $\bar{x} = \dfrac{45 + 48 + 50 + 51}{4} = \dfrac{194}{4} = 48.5$ $R = 51 - 45 = 6$

Sample 2: $\bar{x} = \dfrac{41 + 42 + 43 + 44}{4} = \dfrac{170}{4} = 42.5$ $R = 44 - 41 = 3$

Sample 3: $\bar{x} = \dfrac{46 + 42 + 48 + 42}{4} = \dfrac{178}{4} = 44.5$ $R = 48 - 42 = 6$

Sample 4: $\bar{x} = \dfrac{48 + 46 + 42 + 44}{4} = \dfrac{180}{4} = 45$ $R = 48 - 42 = 6$

Sample 5: $\bar{x} = \dfrac{44 + 43 + 44 + 47}{4} = \dfrac{178}{4} = 44.5$ $R = 47 - 44 = 3$

2. Find $\bar{x}_{GM}$ and $\bar{R}$.

$$\bar{x}_{GM} = \frac{48.5 + 42.5 + 44.5 + 45 + 44.5}{5} = \frac{225}{5} = 45 \qquad \bar{R} = \frac{6 + 3 + 6 + 6 + 3}{5} = \frac{24}{5} = 4.8$$

3. $\text{UCL}_{\bar{x}} = \bar{x}_{GM} + A_2\,\bar{R}$ A_2 is obtained from Table O for n = number in each sample = 4

$\text{UCL}_{\bar{x}} = 45 + .729(4.8) = 48.5$ $\text{LCL}_{\bar{x}} = \bar{x}_{GM} - A_2\,\bar{R} = 45 - .729(4.8) = 41.5$

4.

```
     |_ _ _ _ _ _ _ _ _ _ _ _ _ _ _ _ _ _ _ _48. 5 UCL x̄
  48 +
     |
  47 +
     |
  46 +
     |_____ 45 x̄GM
  44 +
     |
  43 +
     |
  42 +_ _ _ _ _ _ _ _ _ _ _ _ _ _ _ _ _ _ _ _ _41.5 LCL x̄
```

5. Plot the mean of each sample on the chart and connect the points with a line.

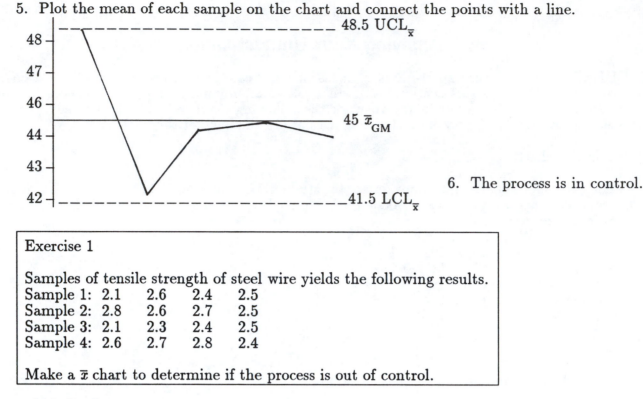

6. The process is in control.

Exercise 1

Samples of tensile strength of steel wire yields the following results.
Sample 1: 2.1   2.6   2.4   2.5
Sample 2: 2.8   2.6   2.7   2.5
Sample 3: 2.1   2.3   2.4   2.5
Sample 4: 2.6   2.7   2.8   2.4

Make a $\bar{x}$ chart to determine if the process is out of control.

Example 2--R chart.

Make a R chart for the data in Example 1.

Solution

1. Find the range for each sample.
    Sample 1: $R = 51 - 45 = 6$
    Sample 2: $R = 44 - 41 = 3$
    Sample 3: $R = 48 - 42 = 6$
    Sample 4: $R = 48 - 42 = 6$
    Sample 5: $R = 47 - 44 = 3$

2. $\bar{R} = \dfrac{6 + 3 + 6 + 6 + 3}{5} = \dfrac{24}{5} = 4.8$

3. $UCL_R = D_4\,\bar{R}$ D_4 is obtained from Table O for $n = $ number in each sample $= 4$

    $UCL_R = 2.115(4.8) = 10.152$ $LCL_R = D_3\,\bar{R} = 0(4.8) = 0$

4.

5.   Plot the range of each sample on the chart and connect the points with a line.

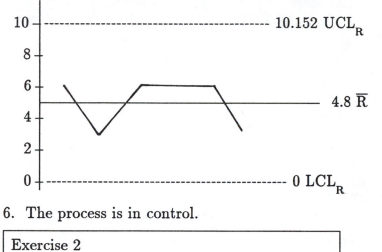

6.   The process is in control.

Exercise 2

Make a R chart for the data in Exercise 1.

Example 3--$\bar{p}$ chart.

Make a control chart for the following number of defectives.

	Parts	Defectives
Sample 1:	200	5
Sample 2:	200	10
Sample 3:	200	6
Sample 4:	200	3

Solution

1.   Find the proportion of defectives for each sample.

Sample 1: $p = \frac{5}{200} = .025$ Sample 2: $p = \frac{10}{200} = .05$

Sample 2: $p = \frac{6}{200} = .03$ Sample 4: $p = \frac{3}{200} = .015$

2. $\bar{p} = \dfrac{.025 + .05 + .03 + .015}{4} = \dfrac{.12}{4} = .03$

3. $UCL_p = \bar{p} + 3\sqrt{\dfrac{\bar{p}(1-\bar{p})}{n}} = .03 + 3\sqrt{\dfrac{.03(.97)}{200}} = .066$

$LCL_p = \bar{p} - 3\sqrt{\dfrac{\bar{p}(1-\bar{p})}{n}} = .03 - 3\sqrt{\dfrac{.03(.97)}{200}} = -.006 = 0$ (can not be less than 0)

4.
```
.066 UCL_p


.03 p̄


0 LCL_p
```

5. Plot the proportion of each sample on the chart and connect the points with a line.

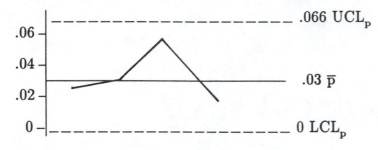

6. The process is in control.

Exercise 3

A service manager of a copier company asks samples of 100 customers if their copiers are working satisfactorily.

	Customers	Dissatisfied
Sample 1:	100	8
Sample 2:	100	5
Sample 3:	100	6
Sample 4:	100	3
Sample 5:	100	2

Make a $\bar{p}$ chart to determine if the copiers are in control.

Example 4--$\bar{c}$ chart.

Ten pages of a manuscript that had errors were studied and the number of errors per page are listed below. Make a $\bar{c}$ of the errors.

4   3   1   1   2   6   2   1   1   2

Solution

1. $\bar{c} = \dfrac{4 + 3 + 1 + 1 + 2 + 6 + 2 + 1 + 1 + 2}{10} = \dfrac{23}{10} = 2.3$

2. $UCL_c = \bar{c} + 3\sqrt{\bar{c}} = 2.3 + 3\sqrt{2.3} = 6.85$

   $LCL_c = \bar{c} - 3\sqrt{\bar{c}} = 2.3 - 3\sqrt{2.3} = -2.2 = 0$

3.

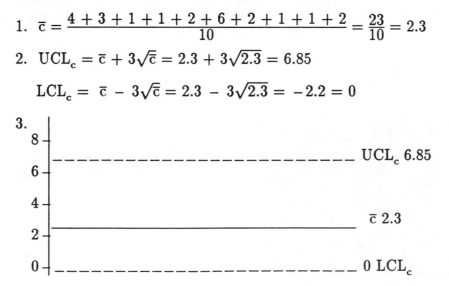

4. Plot the number of defectives in each sample on the chart and connect the points with a line.

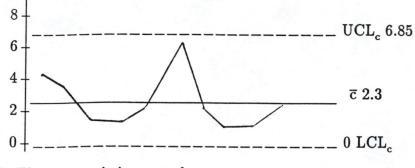

5. The process is in control.

Exercise 4

Make a $\bar{c}$ chart for the number of flaws on a car door that has been repainted.
4   3   5   6   8   7   5   10

QUESTIONS:

QUESTIONS:

Practice Test

1. Make a $\bar{x}$ and a R chart.
   Diameters of bolts   cm

Sample 1:	1.0	0.9	1.3	1.2
Sample 2:	1.5	1.6	1.4	1.5
Sample 3:	1.4	1.6	1.7	1.5
Sample 4:	1.4	1.4	1.3	1.6

2. Make a $\bar{p}$ chart for the number of defectives in samples of 10,000 condoms.

	Condoms	Defectives
Sample 1:	10,000	25
Sample 2:	10,000	18
Sample 3:	10,000	24
Sample 4:	10,000	22
Sample 5:	10,000	40
Sample 6:	10,000	21

3. Make a $\bar{c}$ chart for the number of defectives per part for these defective parts.

    3   7   1   8   4   4   2   2   6   1   8   5

IMPORTANT FORMULAS

MEAN

Sample

$$\bar{x} = \frac{\Sigma x}{n}$$

n = number of values

Population

$$\mu = \frac{\Sigma x}{N}$$

N = number of values

Frequency Distribution

$$\bar{x} = \frac{\Sigma fx}{n}$$

$n = \Sigma f$

Median

Sample or Population
MD = middle number or the average
of the middle two numbers
when the numbers are put in order

Frequency Distribution

$$MD = \frac{\frac{n}{2} - CF}{f}(w) + L_m$$

Variance and Standard Deviation

Sample

$$s^2 = \frac{\Sigma(x - \bar{x})^2}{n - 1} \quad \text{or}$$

$$s^2 = \frac{\Sigma x^2 - \frac{(\Sigma x)^2}{n}}{n - 1}$$

$$s = \sqrt{\frac{\Sigma(x - \bar{x})^2}{n - 1}} \quad \text{or}$$

$$s = \sqrt{\frac{\Sigma x^2 - \frac{(\Sigma x)^2}{n}}{n - 1}}$$

Population

$$\sigma^2 = \frac{\Sigma(x - \mu)^2}{N} \quad \text{or}$$

$$\sigma^2 = \frac{\Sigma x^2 - \frac{(\Sigma x)^2}{N}}{N}$$

$$\sigma = \sqrt{\frac{\Sigma(x - \mu)^2}{N}} \quad \text{or}$$

$$s = \sqrt{\frac{\Sigma x^2 - \frac{(\Sigma x)^2}{N}}{N}}$$

Frequency Distribution

$$s^2 = \frac{\Sigma fx^2 - \frac{(\Sigma fx)^2}{n}}{n - 1}$$

$$s\sqrt{\frac{\Sigma fx^2 - \frac{(\Sigma fx)^2}{n}}{n - 1}}$$

Coefficient of Variation

Sample

$$CV = \frac{s}{\bar{x}}(100\%)$$

Population

$$CV = \frac{\sigma}{\mu}(100\%)$$

Z Score

Sample

$$z = \frac{x - \bar{x}}{s}$$

Population

$$z = \frac{x - \mu}{\sigma}$$

Counting

Permutations (arrangements, no repetitions)

$$_nP_r = \frac{n!}{(n - r)!}$$

n objects taken r at a time

Combination (groups, no repetition)

$$_nC_r = \frac{n!}{(n - r)!r!}$$

n objects taken r at a time

Probability

P(A or B) = P(A) + P(B) if A and B are mutually exclusive
P(A or B) = P(A) + P(B) − P(A and B) if A and B are not mutually exclusive

P(A and B) = P(A)·P(B) if A and B are independent
P(A and B) = P(A)·P(B/A) if A and B are not independent

$P(\bar{A}) = 1 - P(A)$

Probability Distribution

Mean   Standard Deviation   Variation

$$\mu = \Sigma x P(x) \qquad\qquad \sigma^2 = \Sigma x^2 P(x) - \mu^2 \qquad\qquad \sigma = \sqrt{\Sigma x^2 P(x) - \mu^2}$$

Binomial Distribution

$$P(x) = \frac{n!}{(n-x)!x!} \, p^x q^{n-x} \qquad\qquad \mu = np \qquad \sigma^2 = npq \quad \sigma = \sqrt{npq}$$

Multinomial Distribution   ## Hypergeometric Distribution

$$P(M) = \frac{n!}{x_1! x_2! \cdots x_k!} \, p_1^{x_1} p_2^{x_2} \cdots p_k^{x_k} \qquad\qquad P(A) = \frac{{}_aC_x \, {}_bC_{n-x}}{{}_{(a+b)}C_n}$$

Confidence Intervals and Minimum Sample Sizes

Confidence Interval for Population Mean
σ known or $n \geq 30$ σ unknown and $n < 30$

$$\bar{x} - z\left(\frac{\sigma}{\sqrt{n}}\right) \leq \mu \leq \bar{x} + z\left(\frac{\sigma}{\sqrt{n}}\right) \qquad\qquad \bar{x} - t\left(\frac{s}{\sqrt{n}}\right) \leq \mu \leq \bar{x} + t\left(\frac{s}{\sqrt{n}}\right) \qquad df = n - 1$$

Confidence Interval for Population Proportion   Confidence Interval for Population Variance
$n\hat{p}$ and $n\hat{q} \geq 5$

$$\hat{p} - z\sqrt{\frac{\hat{p}\hat{q}}{n}} \leq p \leq \hat{p} + z\sqrt{\frac{\hat{p}\hat{q}}{n}} \qquad\qquad \frac{(n-1)s^2}{\chi^2_{larger}} \leq \hat{p} \leq \frac{(n-1)s^2}{\chi^2_{smaller}}$$

Minimum sample size to estimate a mean   Minimum sample size to estimate a proportion

$$n = \frac{z\sigma}{E} \qquad\qquad\qquad\qquad n = \hat{p}\hat{q}\left(\frac{z}{E}\right)^2$$

Correlation Analysis

$$r = \frac{n\Sigma xy - (\Sigma x)(\Sigma y)}{\sqrt{\left[n\Sigma x^2 - (\Sigma x)^2\right]\left[n\Sigma y^2 - (\Sigma y)^2\right]}}$$

Equation: $y' = ax + b$

$$a = \frac{\Sigma x \Sigma x^2 - \Sigma x \Sigma xy}{n\Sigma x^2 - (\Sigma x)^2} \qquad\qquad b = \frac{n\Sigma xy - \Sigma x \Sigma y}{n\Sigma x^2 - (\Sigma x)^2}$$

$$s_{est} = \sqrt{\frac{\Sigma(y - y')^2}{n - 2}} \qquad \text{or} \qquad \sqrt{\frac{\Sigma y^2 - a\Sigma y - b\Sigma xy}{n - 2}}$$

SUMMARY OF HYPOTHESIS TESTS

If the populations are normally distributed:

Test Means

I. One Sample Mean

   A. σ known or $n \geq 30$, z test

$$z = \frac{\bar{x} - \mu}{\frac{\sigma}{\sqrt{n}}}$$

   B. σ unknown and $n < 30$, t test

$$t = \frac{\bar{x} - \mu}{\frac{s}{\sqrt{n}}} \qquad \mathrm{df} = n - 1$$

II. Two Samples Means

   A. Samples are Independent $(\mu_1 - \mu_2)$ is assumed to be 0 unless stated otherwise

      1. σ known or both n_1 and $n^2 \geq 30$, z test

$$z = \frac{(\bar{x}_1 - \bar{x}_2) - (\mu_1 - \mu_2)}{\sqrt{\frac{\sigma_1^2}{n_1} + \frac{\sigma_2^2}{n_2}}}$$

      2. σ unknown and either n_1 or $n^2 < 30$, t test

        a. if the variances are unequal
          df = smaller of $n_1 - 1$ or $n_2 - 1$

$$t = \frac{(\bar{x}_1 - \bar{x}_2) - (\mu_1 - \mu_2)}{\sqrt{\frac{s_1^2}{n_1} + \frac{s_2^2}{n_2}}}$$

        b. if the variances are equal
          df = $n_1 + n_2 - 2$

$$t = \frac{(\bar{x}_1 - \bar{x}_2) - (\mu_1 - \mu_2)}{\sqrt{\frac{(n_1 - 1)s_1^2 + (n_2 - 1)s_2^2}{n_1 + n_2 - 2}} \sqrt{\frac{1}{n_1} + \frac{1}{n_2}}}$$

   B. Samples are Dependent, t test for differences

$$t = \frac{\bar{D} - \mu_D}{\frac{s_D}{\sqrt{n}}} \qquad \bar{D} = \frac{\Sigma D}{n} \qquad s_D = \sqrt{\frac{\Sigma D^2 - \frac{(\Sigma D)^2}{n}}{n - 1}}$$

III. Three or More Sample Means

   A. Test for differences, F test

$$F = \frac{s_B^2}{s_W^2} \qquad s_B^2 = \frac{\Sigma n(\bar{x} - \bar{x}_{GM})}{k - 1} \qquad s_W^2 = \frac{\Sigma(n - 1)s^2}{\Sigma(n - 1)} \qquad \bar{x}_{GM} = \frac{\Sigma \bar{x}}{n}$$

   B. Test to see where the differences are (if there is a difference), Scheffé test

$$F_s = \frac{\Sigma(\bar{x}_1 - \bar{x}_2)^2}{s_W^2 \left(\frac{1}{n_1} + \frac{1}{n_2}\right)}$$

      dfN = $k - 1$ dfD = $N - k$
      Critical value $F' = (k - 1)$(table value)

Test Proportions

I. One sample proportion, z test

np and nq must be greater than or equal to 5

$$z = \frac{x - np}{\sqrt{npq}}$$

II. Two sample proportions, z test

$$z = \frac{\hat{p}_1 - \hat{p}_2}{\sqrt{\bar{p}\,\bar{q}\left(\frac{1}{n_1} + \frac{1}{n_2}\right)}} \qquad \hat{p}_1 = \frac{x_1}{n_1} \qquad \bar{p} = \frac{x_1 + x_2}{n_1 + n_2} \qquad \bar{q} = 1 - \bar{p}$$

Test Variances

I. One sample variance, χ^2 test

$$\chi^2 = \frac{(n - 1)s^2}{\sigma^2} \qquad df = n - 1$$

II. Two sample variances, F test

$$F = \frac{s_1^2}{s_2^2} \qquad \text{where } s_1^2 \text{ is the larger variance} \qquad dfN = n_1 - 1 \qquad dfD = n_2 - 1$$

Test Frequency Distributions

I. Frequency distributions for two groups if each observed frequency ≥ 5, χ^2 test

$$\chi^2 = \frac{\sum(O - E)^2}{E} \qquad df = \text{number of categories} - 1$$

II. Test two frequency distribution for independence, χ^2 test

$$\chi^2 = \frac{\sum(O - E)^2}{E} \qquad E = \frac{(\text{Row sum})(\text{Column sum})}{\text{Grand Total}} \qquad df = (R - 1)(C - 1)$$

Test Medians

I. $n \leq 25$, single sample sign test

II. $n > 25$, z test

$$z = \frac{(x + 0.5) - \frac{n}{2}}{\frac{\sqrt{n}}{2}}$$

If populations are not normally distributed:

Test means

I. Two dependent samples, paired sample sign test

II. Two dependent samples, Wilcoxon rank sum test

$$z = \frac{R - \mu_R}{\sigma_R} \qquad \mu_R = \frac{n_2(n_1 + n_2 + 1)}{2} \qquad \sigma_R = \sqrt{\frac{n_1 n_2(n_1 + n_2 + 1)}{12}}$$

III. Three or more means if each sample has 5 or more numbers, Kruskal-Wallis or H test

$$H = \frac{12}{n(N + 1)}\left[\frac{R_1^2}{n_1} + \frac{R_2^2}{n_2} + \cdots + \frac{R_k^2}{n_k}\right] - 3(N + 1)$$

Test correlations

Spearman rank correlation coefficient

$$r = 1 - \frac{6\sum d^2}{n(n^2 - 1)} \qquad d = \text{rank of first group minus the rank of the second group}$$

Test randomness, test for runs

ANSWERS

Chapter 1 Checking

I.   Descriptive statistics summarizes or organizes data, but inferential statistics makes comparisons or inferences from the data.

II.   Discrete variables are obtained by counting and will be whole numbers. Continuous variables are obtained by measuring and can take on any value between two amounts.

III.   1. b   2. d   3. a   4. b   5. a   6. c   7. c   8. c   9. d   10. d   11. a

Chapter 2 Checking

I.   Too few classes (should have between 5 and 15 classes), overlapping classes, not continuous (gap between 30 and 40), unequal intervals.

II.   Bars should all be the same width and the same distance apart. Labels on the side do not represent equal amounts.

III.   1. b   2. c   3. c   4. d

Chapter 2 Applying

1. Age of students (one possibility)

Boundaries	Age	f	cf
16.5 - 19.5	17 - 19	6	6
19.5 - 22.5	20 - 22	4	10
22.5 - 25.5	23 - 25	6	16
25.5 - 28.5	26 - 28	6	22
28.5 - 31.5	29 - 31	4	26
31.5 - 34.5	32 - 34	2	28
35.5 - 37.5	35 - 37	2	30
		30	

2. One possible solution

Boundaries	GPA	f	cf
1.495 - 1.995	1.50 - 1.99	4	4
1.995 - 2.495	2.00 - 2.49	4	8
2.495 - 2.995	2.50 - 2.99	4	12
2.995 - 3.495	3.00 - 3.49	2	14
3.495 - 3.995	3.50 - 3.99	2	16
		16	

3. College Algebra Grades

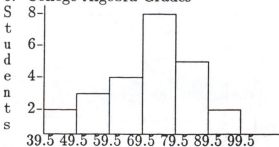

4. College Algebra Grades

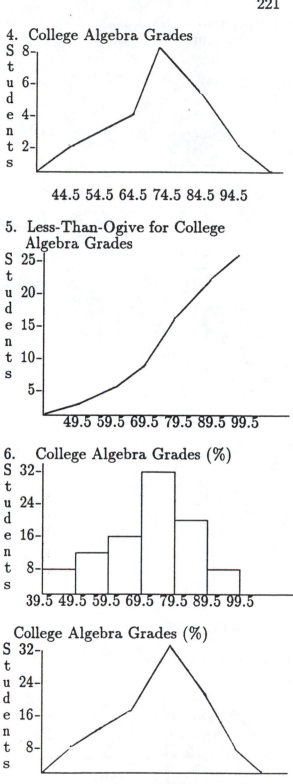

5. Less-Than-Ogive for College Algebra Grades

6. College Algebra Grades (%)

College Algebra Grades (%)

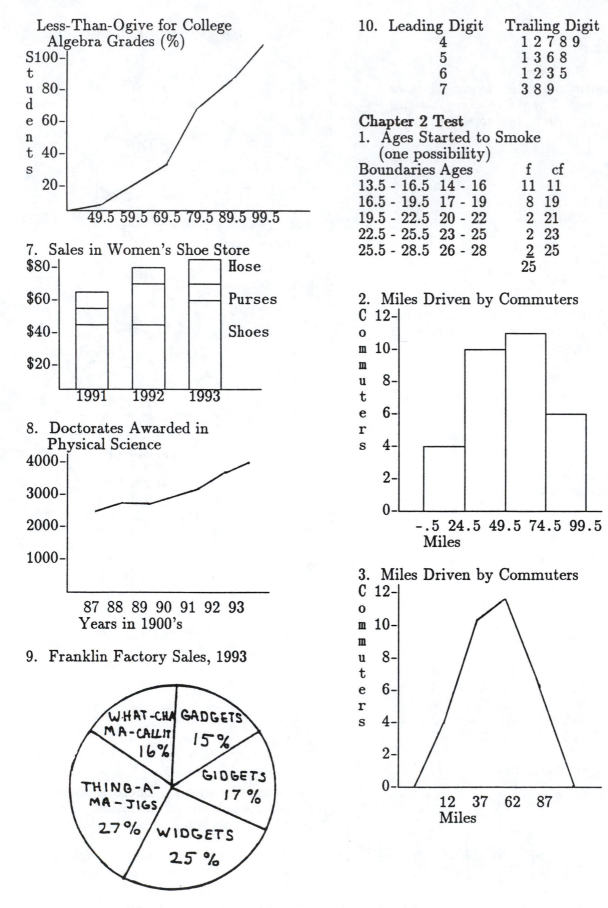

Less-Than-Ogive for College Algebra Grades (%)

(y-axis: Students, values 20, 40, 60, 80, 100; x-axis: 49.5 59.5 69.5 79.5 89.5 99.5)

10.
Leading Digit	Trailing Digit
4	1 2 7 8 9
5	1 3 6 8
6	1 2 3 5
7	3 8 9

Chapter 2 Test

1. Ages Started to Smoke (one possibility)

Boundaries	Ages	f	cf
13.5 - 16.5	14 - 16	11	11
16.5 - 19.5	17 - 19	8	19
19.5 - 22.5	20 - 22	2	21
22.5 - 25.5	23 - 25	2	23
25.5 - 28.5	26 - 28	2	25
		25	

7. **Sales in Women's Shoe Store**

(y-axis: $20, $40, $60, $80; x-axis: 1991 1992 1993; stacked bars labeled Hose, Purses, Shoes)

8. **Doctorates Awarded in Physical Science**

(y-axis: 1000, 2000, 3000, 4000; x-axis: 87 88 89 90 91 92 93, Years in 1900's)

2. **Miles Driven by Commuters**

(y-axis: Commuters, values 0, 2, 4, 6, 8, 10, 12; x-axis: -.5 24.5 49.5 74.5 99.5, Miles)

3. **Miles Driven by Commuters**

(y-axis: Commuters, values 0, 2, 4, 6, 8, 10, 12; x-axis: 12 37 62 87, Miles)

9. **Franklin Factory Sales, 1993**

(pie chart: WHAT-CHA-MA-CALLIT 16%, GADGETS 15%, GIDGETS 17%, WIDGETS 25%, THING-A-MA-JIGS 27%)

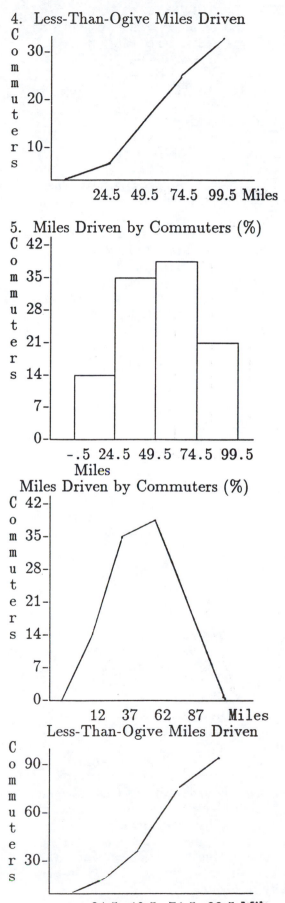

4. Less-Than-Ogive Miles Driven

5. Miles Driven by Commuters (%)

Miles Driven by Commuters (%)

Less-Than-Ogive Miles Driven

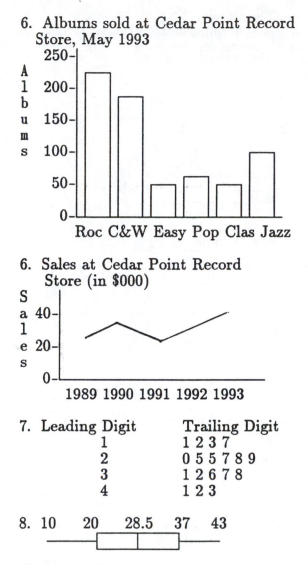

6. Albums sold at Cedar Point Record Store, May 1993

6. Sales at Cedar Point Record Store (in $000)

7.
Leading Digit	Trailing Digit
1	1 2 3 7
2	0 5 5 7 8 9
3	1 2 6 7 8
4	1 2 3

8. 10   20   28.5   37   43

Chapter 3 Checking

I. The median, 280, would probably be the best representation. Since the mean is quite a bit larger than the median, the mean has been affected by a few large numbers.

II. Half of the prices were less than $2.68 and half of the prices were more than $2.68.

III. The grades were an average of 6.2 away from the mean. There is some, but not a whole lot of deviation. The test grades were fairly close together.

IV. The second class had a slightly higher average, but since the standard deviation was higher for the second class, the grades were not as close together for the second class.

V. 72% means the student answered 72% of the questions correctly. 85th percentile means the student did better than 85% of the other students.

VI. 1. c   2. b   3. c   4. a   5.a

Chapter 3 Applying
1. $\bar{x} = \$7.83$ 2. $\bar{x} = 9.6$ 3. $\bar{x} = 69.8$
4. MD = 7.95   5. MD = 5.5   6. 10
7. 71.6   8. no mode   9. modes = 3,6
10. mode = 9   11. mode = 73
12. 5.5   13. 7   14. s = 2.3
15. s = 15.4   16. The first set
17. 77.3%   18. 1832.48 and 2919.52
19. 3 and 33
20. P = 4.5, 75th percentile
21. $P_{20} = 50$, 78th percentile
22.

```
2    3          5.5      7          9
   ┌─────────┬──────────┐
───┤         │          ├────────────
   └─────────┴──────────┘
2    3    4    5    6    7    8    9
```

Chapter 3 Test
1. $\bar{x} = 5$, MD = 5, mode = 4, midrange = 5.5 The mean or average number of pages per article was 5. Half of the articles had more than 5 pages and half had less than 5 pages. The number of pages that appeared the most often was 4.
2. $\mu = .6$ days, MD = 0, mode = 0
3. $\bar{x} = 51.1$, MD = 51.8, mode = 62 The commuters drove an average of 51.1 miles a day and half drove less than 51.8 miles a day. The number of miles drove by the most commuters was 62.
4. $s = 1.4$ This is a fairly small deviation. There was not a lot of variation in the number of pages in the magazine articles.
5. $s = 23.4$ The numbers are an average of 23.4 away from the mean of 51.2. This is a large deviation.
6. The pills from plant A have an average that is further from the desired 50 mg than the pills from plant B. The CV of A = 4.4% and for B, CV = 17%. There is more deviation in the pills from plant B. The pills from plant B are not as uniform.
7. VCR: $z = -1.46$ TV: $z = -1.35$ The VCR was slightly more below its average selling price than the TV was, but there is very little difference.

8. 75%   9. 210-258
10a. P = 85   b. P = 30   c. 60th
11a. 15   b. 78th percentile
12. 14   16 17   20   26

```
   14  16 17        20              26
   ┌───────┬──────────────┐
───┤       │              ├──────────
   └───────┴──────────────┘
   14   16   18   20   22   24   26
```

Chapter 4 Checking
Ia. This is a permutation because the order matters. The first one bats first and so on.
Ib. This is a combination because the order the person is chosen does not matter. He will be in the group whether he is chosen first or second.
Ic. This is a combination because the five girls are in the group no matter what order they are picked in.
Id. This is a permutation since the order makes the difference of who gets first, second, or third place.
Ie. This a combination. Order does not matter since they all get the same amount.
If. This is a combination since the order does not matter. Nebraska versus Oklahoma is the same game as Oklahoma versus Nebraska.
II. 1. b   2. a   3. a   4. b

Chapter 4 Applying
1. 120   2. 24   3. 6   4. 4080
5. 2,522,520   6. 120   7. 560

Chapter 4 Test
1. 1024   2. 1440   3. 95,040
4. 17,576   5. 90720   6. 28
7. 194,040   8. 270

Chapter 5 Checking
Ia. Mutually exclusive events - two events that can not happen at the same time.
Ib. Independent events - two events where the outcome of one event has no effect on the outcome of the second event.
Ic. Complement - all the outcomes not in the event.
II. Classical probability is used when each event has the same chance of happening. In empirical probability an experiment has to be performed to get the relative frequencies since the events can not be assumed to be equally to occur.

III. The rules of addition are used when the probability of one event over another event is desired. The first rule is used if the events are mutually exclusive. If the events are not mutually exclusive, the second rule says to subtract the common event from the total.

IV. The rules of multiplication are used when looking for the probability of one event and another event. The first rule is used if the events are independent. If the outcome of the first event has an effect on the probability of the second event, then the first event is assumed to have happened when you find the probability of the second event.

V. 1. c   2. c   3. a   4. d   5. b   6. b
7. c   8. d   9. c   10. b   11. a   12. c

Chapter 5 Applying
1. $\frac{2}{9}$ 3. $\frac{18}{41}$ 3. $\frac{29}{109}$ 4. $\frac{11}{15}$
5. .0165   6. 51%   7. $\frac{1}{406}$
8. .53   9. $\frac{1}{6}$ 10. .488
11. .995   12. .201   13. .000061

Chapter 5 Test
1a. $\frac{1}{13}$ b. $\frac{1}{2}$ c. $\frac{7}{13}$
2a. .372   b. .285   c. .715
3. .7   4. .202   5. .58
6. .0122   7. .075   8a. $\frac{4}{11}$ b. $\frac{23}{44}$
9. .478   10. .324   11. .625

Chapter 6 Checking
Ia. yes   b. yes   c. No, one of the probabilities is negative.   d. No, sum of the probabilities is not 1.
II. 1. c   2. a   3. d   4. b   5. a

Chapter 6 Applying
1. $E(x) = .5$ $\sigma = 22.36$ 2. .075
3. .323   4. $\mu = 28$ $\sigma = 3.5$
5. .0083   6. .1044   7. .1462
8. .1637   9. .196

Chapter 6 Test
1. $\mu = 6.78$ $\sigma = 2.93$ 2. .215
3. .930   4. $\mu = 294$ $\sigma = 2.4$
5. .082   6. .0286   7. .0653
8. .0007   9. .282

Chapter 7 Checking
I. The mean, the median, and the mode must be all equal.
II. np and nq are both $\geq$ 5.
III. 1. b   2. a   3. c   4. b

Chapter 7 Applying
1a. .3944   b. .1056   c. .8944
2a. .0301   b. .8708   c. .0991   d. .9686
3. 5   4. 42.4   5. 31.8   6. .0084
7. .9251

Chapter 7 Test
1a. .4525   b. .9525   c. .0475
d. .2454   e. .2546   f. .7454
g. .6979   h. .2450
2. 232.4   3. .0162   4. 131
5. .0089   6. .0918

Chapter 8 Checking
I. 1. sample mean   2. sample proportion   3. .05   4. two
5. maximum error of estimate

Chapter 8 Applying
1. $81.98 \leq \mu \leq 84.02$
2. $1445.76 \leq \mu \leq 1478.24$
3. 25   4. $.142 \leq p \leq .278$ 5. 423

Chapter 8 Test
1. $20.8 \leq \mu \leq 22.4$
2. $140 \leq \mu \leq 152$ 3. 14
4. $.145 \leq p \leq .275$ 5. 972

Chapter 9 Checking
I. $H_0: \mu \leq 32$ $H_1: \mu > 32$
II. $H_0: \mu \leq 17$ $H_1: \mu > 17$
III. $H_0: \mu \geq 12.3$ $H_1: \mu < 12.3$
IV. $H_0: \mu = 18$ $H_1: \mu \neq 18$
V. 1. a   2. a   3. b   4. c   5. b
6. b   7. a

Chapter 9 Applying
1. Critical value = 2.326   test value = 3.125   Reject the null. The number of items produced is greater than 145.
2. Critical value = ± 2.576, test value = 2.828   Reject the null. The time is not 12 minutes.
3. Critical value = -1.796 test value = -1.540 Accept the null. The response time is not less than 10 minutes.
4. Critical value = ± 2.576 test value = $-.612$ Accept the null. 40% do make a purchase.
5. .0618   6. .0023

Chapter 9 Test
1. Critical value = -1.645 test value = -3.637 Reject the null. The new system does lower the the response time.
2. Critical values = ± 2.576 test value = -2.055 Accept the null. The average mpg is not changed.
3. $p = .0394$
4. Critical value = -2.718 test value = -3.849 Reject the null. The machine does not do at least 25,000 copies.
5. Critical value = ± 2.567 test value = $-.943$ Accept the null. 10% do win.

Chapter 10 Checking
Ia. The populations must be normally distributed and the samples must be independent. The population standard deviations must be known or both sample sizes must be greater than or equal to 30.
Ib. The populations must be normally distributed and the samples must be independent. The population standard deviations must be unknown or either sample sizes must be less than 30. Two different tests are used depending on whether the variances of the two samples are assumed to be equal or not.
Ic. The populations must be normally distributed and the samples must be dependent.
II. 1. c   2. d   3. b   4. b   5. c

Chapter 10 Applying
1. critical values = ± 2.576 test value = -1.036 Accept the null. There is no significant difference in the two classes.
2. Critical values = ± 2.306 test value = -2.356 Reject the null. These cars have different gas mileages.
3. Critical values = -2.326 test value = -3.274 Reject the null. Machine A does put in less than Machine B.
4. Critical values = ± 3.250 test value = 2.141 Accept the null. There is no difference in these IQ scores.
5. Critical values = ± 1.960 test value = 2.188 Reject the null. There is a difference in the proportion of people that recycle in large towns and in small towns.

Chapter 10 Test
1. Critical values = 1.645 test value = 4.491 Reject the null. Brand X does have more ibuprofen.
2. Critical values = ± 3.106 test value = -3.025 Accept the null. There is not a difference in the amount of time spent by male clerks and the amount of time spent by female clerks with customers.
3. Critical values = -1.645 test value = -2.095 Reject the null. Smoking did lower the birthweights.
4. Critical values = ± 2.262 test value = -1.91 Accept the null. There is no difference in the final exams.
5. Critical values = ± 1.960 test value = $-.472$ Accept the null. There is no difference in the proportion passing using the two methods.

Chapter 11 Checking
I. 1. Strong negative correlation
2. Fair positive correlation
3. Some negative correlation
4. Strong positive correlation
II. 1. x = score on first test y = final grade
2. x = number of drinks y = dexterity
3. x = car speed y = gas mileage
4. x = high school y = college
5. x = age y = grade
6. x = rainfall y = crop yield
III. 1. a   2. b   3. b   4. c

Chapter 11 Applying
1.

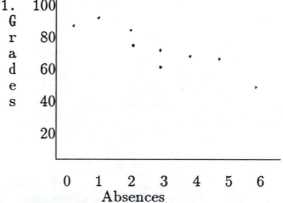

Some negative correlation
2. $r = -.877$ Critical value = .798 There is a strong negative correlation.
3. $r^2 = .768$ 76.8% of the changes in grades can be predicted by changes in absences.

(3) $1 - r^2 = .232$ 23.2% of the changes in grades can not be predicted by the changes in absences.
4. $y' = 89.6 - 5.2x$ 5. $s_{est} = 5.89$
6. 13.05 to 18.15

Chapter 11 Test

1a.

Number of drinks

Some negative correlation
b. $r = -.959$ There is a large negative correlation.
c. $r^2 = .92$ $1 - r^2 = .08$ 92% of the changes in dexterity can be determine by the change in the number of drinks.
d. $y' = 19.25 - 2.92x$ e. $s_{est} = 1.72$
2. -1522.2 to -1345.8

Chapter 12 Checking

I.   If the sample is from a population that is normally distributed, the test for a single variance is used to compare the sample variance to the population variances.
II.   The test for goodness of fit is used to see if a frequency distribution fits a specific pattern and can be used if the expected frequencies are at least 5 for each group.
III.   Tests for independence are used to see if there is a difference in two different groups.

Chapter 12 Applying

1.   Critical values = 14.573 and 43.194 test value = 25.393   Accept the null. The variance is 84.
2.   Critical value = 22.164   test value = 26.615   Accept the null.   The variance is 125.
3. $40.2 \leq \sigma^2 \leq 238.5$
4.   Critical value = 9.488 test value = 15.68   Reject the null.   There is a difference in the number of absences per day.
5.   Critical value = 5.991 test value = 14.06 Reject the null.   The enrollments do not fit the pattern.

6.   Critical value = 6.635   test value = .58   Accept the null.   Number of smokers is independent from sex.

Chapter 12 Test

1.   Critical value = 2.088 test value = 9.342 Accept the null.   The variance is greater than or equal to 158.
2. $201 \leq \sigma^2 \leq 866$
3.   Critical value = 9.488 test value = 31.4   Reject the null.   There is a preference.
4.   Critical value = 7.815 test value = 2.66   Accept the null.   The grade is independent from sex.

Chapter 13 Checking

I.   If the samples are independent and from normally distributed populations, the F test can be used to compare two variances.
II.   The F test for analysis of variance (ANOVA) can be used to see of there is a difference among three or more means if the samples are from normally distributed populations, the samples are independent, and the variances of the populations are equal.
III.   The Scheffé test is used to compare two means at a time where the difference is if ANOVA shows that there is a difference among the means.

Chapter 13 Applying

1.   Critical value = 4.40   test value = 3.58 Accept the null.   The variance for Brand A is not greater than for Brand B.
2.   Critical value = 2.86   test value = 3.568   Reject the null. There is a difference in the variances in cholesterol levels.
3. $\bar{x}_1 = 25.6$ $s_1^2 = 153.3$ $n_1 = 5$
$\bar{x}_2 = 27.5$ $s_2^2 = 101.0$ $n_2 = 6$
$\bar{x}_3 = 20.43$ $s_3^2 = 92.9$ $n_3 = 7$
Critical value = 6.36   test value = .788   There is a no difference among the three different locations.

Chapter 13 Test

1.   Critical value = 2.47   test value = 2.49   Reject the null.   The variances are not equal.
2. $\bar{x}_1 = 4.8$ $s_1^2 = 5.37$ $n_1 = 6$
$\bar{x}_2 = 9.7$ $s_2^2 = 4.33$ $n_2 = 3$
$\bar{x}_3 = 3.2$ $s_3^2 = 1.37$ $n_3 = 6$

Critical value = 6.93   test value = 13.36 There is a difference among the means.

Chapter 14 Checking
1. true   2. false   3. true   4. false
5. true   6. true   7. false   8. true
9. true

Chapter 14 Applying
1. test value = 2 critical value = 3
Reject the null. The median is not 112 for these IQ scores.
2. test value = 3   critical value = 3
Reject the null. The stress levels are lower on Monday.
3. Critical values = ±1.96 test value = −1.29 Accept the null. There is no difference in the amount spent by the customers with credit cards and cash.
4. Critical value = 9.210 test value = 11.92 Reject the null. There is a difference among the amounts produced by the different shifts.
5. Critical value = .786 test value = .857 There is a correlation.
6. Critical values = 11 and 24 test value = 25 Reject the null. This is not random.

Chapter 14 Test
I. Critical value = 2   test value = 6 Accept the null. The median age is 17.
2. Critical value = 1   test value = 1 Reject the null. There is a difference in weight loss with exercise.
3. Critical values = ±2.576 test value = 1.97 There is no difference in the age that men and women start to smoke.
4. Critical value = 9.210 test value = 8.15 There is no difference.
5. Critical value = .866 (if .05 is used) test value = .94 There is a correlation.
6. Critical values = 6 and 16 test value = 11 Accept the null. They are random.

Chapter 15 Checking
Ia. First decide on how many cookies need to be tested. You might decide to test 100 cookies. Look on a random number chart and write down 100 numbers, then check the cookies that match these numbers. This method would get a good sample to see if the machine is working properly, but if there is a large variation in the sample, it might be hard to tell what is wrong.

Ib. If you want to test 100 cookies, you would need to test every 100th cookie. Start with any cookie and test every 100th one. This might not be a very good method for this situation since the cookies are not randomly mixed, but are in the order that they were produced. If there is a problem with the machine this method might not find the problem. For instance, if the first cookies pressed in each batch were too big, this method might not test any of the cookies from the first of a batch.
Ic. Stratified sampling might be used here to see if there is a problem with certain groups of cookies. The cookies for each batch might be divided into those from the first half and those from the the second half and an equal amount taken from each group.
Id. Cluster sampling might be used here by taking one or two complete batches to test.
Ie. Sequence sampling could be used here to test the weight of each cookie and the sequence it was produced to test the uniformity of the batch.

II. 1. c   2. c   3. c   4. d   5. d
6. b   7. a

Chapter 15 Applying and Test
Answers will vary here since different samples could be picked.

Chapter 16 Checking
1. true   2. false   3. true   4. false

Chapter 16 Applying
1.
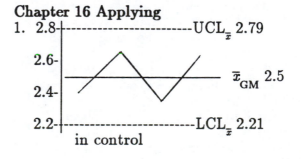
in control

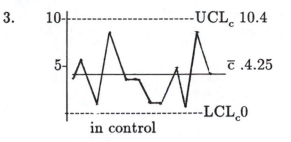

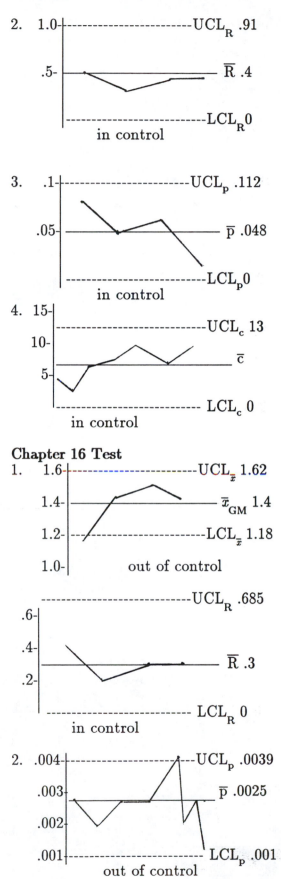

2.   1.0┤----------------UCL_R .91

     .5┤   R̄ .4

        ----------------LCL_R 0
              in control

3.   .1┤----------------UCL_p .112

     .05┤   p̄ .048

        ----------------LCL_p 0
              in control

4. 15┤
    10┤   ----------------UCL_c 13

     5┤   c̄

        ----------------LCL_c 0
            in control

3. 10┤----------------UCL_c 10.4

     5┤   c̄ .4.25

        ----------------LCL_c 0
              in control

Chapter 16 Test

1.   1.6┤----------------UCL_x̄ 1.62

     1.4┤   x̄_GM 1.4

     1.2┤----------------LCL_x̄ 1.18

     1.0┤   out of control

        ----------------UCL_R .685

     .6┤

     .4┤   R̄ .3

     .2┤

        ----------------LCL_R 0
              in control

2. .004┤----------------UCL_p .0039

    .003┤   p̄ .0025

    .002┤

    .001┤----------------LCL_p .001
              out of control

230

NOTES:

Tables

A. Factorials

B. Binomial Distributions

C. Poisson Distributions

D. Table of Random Numbers

E. The Standard Normal Distribution

F. The t Distribution

G. The χ^2 Distribution

H. The F Distribution

I. Critical Values for the PPMC

J. Critical Values for the Sign Test

K. Critical Values for the Wilcoxon Sign-Signed Ranked Test

L. Critical Values for the Rank Correlation Coefficient

M. Critical Values for the Number of Runs (N)

O. Factors for Computing Control Limits

Table A

Factorials

n	$n!$
0	1
1	1
2	2
3	6
4	24
5	120
6	720
7	5,040
8	40,320
9	362,880
10	3,628,800
11	39,916,800
12	479,001,600
13	6,227,020,800
14	87,178,291,200
15	1,307,674,368,000
16	20,922,789,888,000
17	355,687,428,096,000
18	6,402,373,705,728,000
19	121,645,100,408,832,000
20	2,432,902,008,176,640,000

Table B

The Binomial Distribution

n	x	\(p\) 0.05	0.1	0.2	0.3	0.4	0.5	0.6	0.7	0.8	0.9	0.95
2	0	0.902	0.810	0.640	0.490	0.360	0.250	0.160	0.090	0.040	0.010	0.002
	1	0.095	0.180	0.320	0.420	0.480	0.500	0.480	0.420	0.320	0.180	0.095
	2	0.002	0.010	0.040	0.090	0.160	0.250	0.360	0.490	0.640	0.810	0.902
3	0	0.857	0.729	0.512	0.343	0.216	0.125	0.064	0.027	0.008	0.001	
	1	0.135	0.243	0.384	0.441	0.432	0.375	0.288	0.189	0.096	0.027	0.007
	2	0.007	0.027	0.096	0.189	0.288	0.375	0.432	0.441	0.384	0.243	0.135
	3		0.001	0.008	0.027	0.064	0.125	0.216	0.343	0.512	0.729	0.857
4	0	0.815	0.656	0.410	0.240	0.130	0.062	0.026	0.008	0.002		
	1	0.171	0.292	0.410	0.412	0.346	0.250	0.154	0.076	0.026	0.004	
	2	0.014	0.049	0.154	0.265	0.346	0.375	0.346	0.265	0.154	0.049	0.014
	3		0.004	0.026	0.076	0.154	0.250	0.346	0.412	0.410	0.292	0.171
	4			0.002	0.008	0.026	0.062	0.130	0.240	0.410	0.656	0.815
5	0	0.774	0.590	0.328	0.168	0.078	0.031	0.010	0.002			
	1	0.204	0.328	0.410	0.360	0.259	0.156	0.077	0.028	0.006		
	2	0.021	0.073	0.205	0.309	0.346	0.312	0.230	0.132	0.051	0.008	0.001
	3	0.001	0.008	0.051	0.132	0.230	0.312	0.346	0.309	0.205	0.073	0.021
	4			0.006	0.028	0.077	0.156	0.259	0.360	0.410	0.328	0.204
	5				0.002	0.010	0.031	0.078	0.168	0.328	0.590	0.774
6	0	0.735	0.531	0.262	0.118	0.047	0.016	0.004	0.001			
	1	0.232	0.354	0.393	0.303	0.187	0.094	0.037	0.010	0.002		
	2	0.031	0.098	0.246	0.324	0.311	0.234	0.138	0.060	0.015	0.001	
	3	0.002	0.015	0.082	0.185	0.276	0.312	0.276	0.185	0.082	0.015	0.002
	4		0.001	0.015	0.060	0.138	0.234	0.311	0.324	0.246	0.098	0.031
	5			0.002	0.010	0.037	0.094	0.187	0.303	0.393	0.354	0.232
	6				0.001	0.004	0.016	0.047	0.118	0.262	0.531	0.735
7	0	0.698	0.478	0.210	0.082	0.028	0.008	0.002				
	1	0.257	0.372	0.367	0.247	0.131	0.055	0.017	0.004			
	2	0.041	0.124	0.275	0.318	0.261	0.164	0.077	0.025	0.004		
	3	0.004	0.023	0.115	0.227	0.290	0.273	0.194	0.097	0.029	0.003	
	4		0.003	0.029	0.097	0.194	0.273	0.290	0.227	0.115	0.023	0.004
	5			0.004	0.025	0.077	0.164	0.261	0.318	0.275	0.124	0.041
	6				0.004	0.017	0.055	0.131	0.247	0.367	0.372	0.257
	7					0.002	0.008	0.028	0.082	0.210	0.478	0.698

Source: John E. Freund, *Modern Elementary Statistics,* 7e, © 1988, pp. 517–521. Reprinted by permission of Prentice-Hall, Inc., Englewood Cliffs, New Jersey.

Note: All values omitted in this table are 0.0005 or less.

Table B

The Binomial Distribution *Continued*

n	x	\(p\) 0.05	0.1	0.2	0.3	0.4	0.5	0.6	0.7	0.8	0.9	0.95
8	0	0.663	0.430	0.168	0.058	0.017	0.004	0.001				
	1	0.279	0.383	0.336	0.198	0.090	0.031	0.008	0.001			
	2	0.051	0.149	0.294	0.296	0.209	0.109	0.041	0.010	0.001		
	3	0.005	0.033	0.147	0.254	0.279	0.219	0.124	0.047	0.009		
	4		0.005	0.046	0.136	0.232	0.273	0.232	0.136	0.046	0.005	
	5			0.009	0.047	0.124	0.219	0.279	0.254	0.147	0.033	0.005
	6			0.001	0.010	0.041	0.109	0.209	0.296	0.294	0.149	0.051
	7				0.001	0.008	0.031	0.090	0.198	0.336	0.383	0.279
	8					0.001	0.004	0.017	0.058	0.168	0.430	0.663
9	0	0.630	0.387	0.134	0.040	0.010	0.002					
	1	0.299	0.387	0.302	0.156	0.060	0.018	0.004				
	2	0.063	0.172	0.302	0.267	0.161	0.070	0.021	0.004			
	3	0.008	0.045	0.176	0.267	0.251	0.164	0.074	0.021	0.003		
	4	0.001	0.007	0.066	0.172	0.251	0.246	0.167	0.074	0.017	0.001	
	5		0.001	0.017	0.074	0.167	0.246	0.251	0.172	0.066	0.007	0.001
	6			0.003	0.021	0.074	0.164	0.251	0.267	0.176	0.045	0.008
	7				0.004	0.021	0.070	0.161	0.267	0.302	0.172	0.063
	8					0.004	0.018	0.060	0.156	0.302	0.387	0.299
	9						0.002	0.010	0.040	0.134	0.387	0.630
10	0	0.599	0.349	0.107	0.028	0.006	0.001					
	1	0.315	0.387	0.268	0.121	0.040	0.010	0.002				
	2	0.075	0.194	0.302	0.233	0.121	0.044	0.011	0.001			
	3	0.010	0.057	0.201	0.267	0.215	0.117	0.042	0.009	0.001		
	4	0.001	0.011	0.088	0.200	0.251	0.205	0.111	0.037	0.006		
	5		0.001	0.026	0.103	0.201	0.246	0.201	0.103	0.026	0.001	
	6			0.006	0.037	0.111	0.205	0.251	0.200	0.088	0.011	0.001
	7			0.001	0.009	0.042	0.117	0.215	0.267	0.201	0.057	0.010
	8				0.001	0.011	0.044	0.121	0.233	0.302	0.194	0.075
	9					0.002	0.010	0.040	0.121	0.268	0.387	0.315
	10						0.001	0.006	0.028	0.107	0.349	0.599

Table B

The Binomial Distribution *Continued*

n	x	0.05	0.1	0.2	0.3	0.4	0.5	0.6	0.7	0.8	0.9	0.95
11	0	0.569	0.314	0.086	0.020	0.004						
	1	0.329	0.384	0.236	0.093	0.027	0.005	0.001				
	2	0.087	0.213	0.295	0.200	0.089	0.027	0.005	0.001			
	3	0.014	0.071	0.221	0.257	0.177	0.081	0.023	0.004			
	4	0.001	0.016	0.111	0.220	0.236	0.161	0.070	0.017	0.002		
	5		0.002	0.039	0.132	0.221	0.226	0.147	0.057	0.010		
	6			0.010	0.057	0.147	0.226	0.221	0.132	0.039	0.002	
	7			0.002	0.017	0.070	0.161	0.236	0.220	0.111	0.016	0.001
	8				0.004	0.023	0.081	0.177	0.257	0.221	0.071	0.014
	9				0.001	0.005	0.027	0.089	0.200	0.295	0.213	0.087
	10					0.001	0.005	0.027	0.093	0.236	0.384	0.329
	11							0.004	0.020	0.086	0.314	0.569
12	0	0.540	0.282	0.069	0.014	0.002						
	1	0.341	0.377	0.206	0.071	0.017	0.003					
	2	0.099	0.230	0.283	0.168	0.064	0.016	0.002				
	3	0.017	0.085	0.236	0.240	0.142	0.054	0.012	0.001			
	4	0.002	0.021	0.133	0.231	0.213	0.121	0.042	0.008	0.001		
	5		0.004	0.053	0.158	0.227	0.193	0.101	0.029	0.003		
	6			0.016	0.079	0.177	0.226	0.177	0.079	0.016		
	7			0.003	0.029	0.101	0.193	0.227	0.158	0.053	0.004	
	8			0.001	0.008	0.042	0.121	0.213	0.231	0.133	0.021	0.002
	9				0.001	0.012	0.054	0.142	0.240	0.236	0.085	0.017
	10					0.002	0.016	0.064	0.168	0.283	0.230	0.099
	11						0.003	0.017	0.071	0.206	0.377	0.341
	12							0.002	0.014	0.069	0.282	0.540
13	0	0.513	0.254	0.055	0.010	0.001						
	1	0.351	0.367	0.179	0.054	0.011	0.002					
	2	0.111	0.245	0.268	0.139	0.045	0.010	0.001				
	3	0.021	0.100	0.246	0.218	0.111	0.035	0.006	0.001			
	4	0.003	0.028	0.154	0.234	0.184	0.087	0.024	0.003			
	5		0.006	0.069	0.180	0.221	0.157	0.066	0.014	0.001		
	6		0.001	0.023	0.103	0.197	0.209	0.131	0.044	0.006		
	7			0.006	0.044	0.131	0.209	0.197	0.103	0.023	0.001	
	8			0.001	0.014	0.066	0.157	0.221	0.180	0.069	0.006	
	9				0.003	0.024	0.087	0.184	0.234	0.154	0.028	0.003
	10				0.001	0.006	0.035	0.111	0.218	0.246	0.100	0.021
	11					0.001	0.010	0.045	0.139	0.268	0.245	0.111
	12						0.002	0.011	0.054	0.179	0.367	0.351
	13							0.001	0.010	0.055	0.254	0.513

Table B

The Binomial Distribution *Continued*

n	x	0.05	0.1	0.2	0.3	0.4	0.5	0.6	0.7	0.8	0.9	0.95
14	0	0.488	0.229	0.044	0.007	0.001						
	1	0.359	0.356	0.154	0.041	0.007	0.001					
	2	0.123	0.257	0.250	0.113	0.032	0.006	0.001				
	3	0.026	0.114	0.250	0.194	0.085	0.022	0.003				
	4	0.004	0.035	0.172	0.229	0.155	0.061	0.014	0.001			
	5		0.008	0.086	0.196	0.207	0.122	0.041	0.007			
	6		0.001	0.032	0.126	0.207	0.183	0.092	0.023	0.002		
	7			0.009	0.062	0.157	0.209	0.157	0.062	0.009		
	8			0.002	0.023	0.092	0.183	0.207	0.126	0.032	0.001	
	9				0.007	0.041	0.122	0.207	0.196	0.086	0.008	
	10				0.001	0.014	0.061	0.155	0.229	0.172	0.035	0.004
	11					0.003	0.022	0.085	0.194	0.250	0.114	0.026
	12					0.001	0.006	0.032	0.113	0.250	0.257	0.123
	13						0.001	0.007	0.041	0.154	0.356	0.359
	14							0.001	0.007	0.044	0.229	0.488
15	0	0.463	0.206	0.035	0.005							
	1	0.366	0.343	0.132	0.031	0.005						
	2	0.135	0.267	0.231	0.092	0.022	0.003					
	3	0.031	0.129	0.250	0.170	0.063	0.014	0.002				
	4	0.005	0.043	0.188	0.219	0.127	0.042	0.007	0.001			
	5	0.001	0.010	0.103	0.206	0.186	0.092	0.024	0.003			
	6		0.002	0.043	0.147	0.207	0.153	0.061	0.012	0.001		
	7			0.014	0.081	0.177	0.196	0.118	0.035	0.003		
	8			0.003	0.035	0.118	0.196	0.177	0.081	0.014		
	9			0.001	0.012	0.061	0.153	0.207	0.147	0.043	0.002	
	10				0.003	0.024	0.092	0.186	0.206	0.103	0.010	0.001
	11				0.001	0.007	0.042	0.127	0.219	0.188	0.043	0.005
	12					0.002	0.014	0.063	0.170	0.250	0.129	0.031
	13						0.003	0.022	0.092	0.231	0.267	0.135
	14							0.005	0.031	0.132	0.343	0.366
	15								0.005	0.035	0.206	0.463

Table B

The Binomial Distribution *Continued*

n	x	0.05	0.1	0.2	0.3	0.4	0.5	0.6	0.7	0.8	0.9	0.95
16	0	0.440	0.185	0.028	0.003							
	1	0.371	0.329	0.113	0.023	0.003						
	2	0.146	0.275	0.211	0.073	0.015	0.002					
	3	0.036	0.142	0.246	0.146	0.047	0.009	0.001				
	4	0.006	0.051	0.200	0.204	0.101	0.028	0.004				
	5	0.001	0.014	0.120	0.210	0.162	0.067	0.014	0.001			
	6		0.003	0.055	0.165	0.198	0.122	0.039	0.006			
	7			0.020	0.101	0.189	0.175	0.084	0.019	0.001		
	8			0.006	0.049	0.142	0.196	0.142	0.049	0.006		
	9			0.001	0.019	0.084	0.175	0.189	0.101	0.020		
	10				0.006	0.039	0.122	0.198	0.165	0.055	0.003	
	11				0.001	0.014	0.067	0.162	0.210	0.120	0.014	0.001
	12					0.004	0.028	0.101	0.204	0.200	0.051	0.006
	13					0.001	0.009	0.047	0.146	0.246	0.142	0.036
	14						0.002	0.015	0.073	0.211	0.275	0.146
	15							0.003	0.023	0.113	0.329	0.371
	16								0.003	0.028	0.185	0.440
17	0	0.418	0.167	0.023	0.002							
	1	0.374	0.315	0.096	0.017	0.002						
	2	0.158	0.280	0.191	0.058	0.010	0.001					
	3	0.041	0.156	0.239	0.125	0.034	0.005					
	4	0.008	0.060	0.209	0.187	0.080	0.018	0.002				
	5	0.001	0.017	0.136	0.208	0.138	0.047	0.008	0.001			
	6		0.004	0.068	0.178	0.184	0.094	0.024	0.003			
	7		0.001	0.027	0.120	0.193	0.148	0.057	0.009			
	8			0.008	0.064	0.161	0.185	0.107	0.028	0.002		
	9			0.002	0.028	0.107	0.185	0.161	0.064	0.008		
	10				0.009	0.057	0.148	0.193	0.120	0.027	0.001	
	11				0.003	0.024	0.094	0.184	0.178	0.068	0.004	
	12				0.001	0.008	0.047	0.138	0.208	0.136	0.017	0.001
	13					0.002	0.018	0.080	0.187	0.209	0.060	0.008
	14						0.005	0.034	0.125	0.239	0.156	0.041
	15						0.001	0.010	0.058	0.191	0.280	0.158
	16							0.002	0.017	0.096	0.315	0.374
	17								0.002	0.023	0.167	0.418

Table B

The Binomial Distribution *Continued*

n	x	p 0.05	0.1	0.2	0.3	0.4	0.5	0.6	0.7	0.8	0.9	0.95
18	0	0.397	0.150	0.018	0.002							
	1	0.376	0.300	0.081	0.013	0.001						
	2	0.168	0.284	0.172	0.046	0.007	0.001					
	3	0.047	0.168	0.230	0.105	0.025	0.003					
	4	0.009	0.070	0.215	0.168	0.061	0.012	0.001				
	5	0.001	0.022	0.151	0.202	0.115	0.033	0.004				
	6		0.005	0.082	0.187	0.166	0.071	0.015	0.001			
	7		0.001	0.035	0.138	0.189	0.121	0.037	0.005			
	8			0.012	0.081	0.173	0.167	0.077	0.015	0.001		
	9			0.003	0.039	0.128	0.185	0.128	0.039	0.003		
	10			0.001	0.015	0.077	0.167	0.173	0.081	0.012		
	11				0.005	0.037	0.121	0.189	0.138	0.035	0.001	
	12				0.001	0.015	0.071	0.166	0.187	0.082	0.005	
	13					0.004	0.033	0.115	0.202	0.151	0.022	0.001
	14					0.001	0.012	0.061	0.168	0.215	0.070	0.009
	15						0.003	0.025	0.105	0.230	0.168	0.047
	16						0.001	0.007	0.046	0.172	0.284	0.168
	17							0.001	0.013	0.081	0.300	0.376
	18								0.002	0.018	0.150	0.397
19	0	0.377	0.135	0.014	0.001							
	1	0.377	0.285	0.068	0.009	0.001						
	2	0.179	0.285	0.154	0.036	0.005						
	3	0.053	0.180	0.218	0.087	0.017	0.002					
	4	0.011	0.080	0.218	0.149	0.047	0.007	0.001				
	5	0.002	0.027	0.164	0.192	0.093	0.022	0.002				
	6		0.007	0.095	0.192	0.145	0.052	0.008	0.001			
	7		0.001	0.044	0.153	0.180	0.096	0.024	0.002			
	8			0.017	0.098	0.180	0.144	0.053	0.008			
	9			0.005	0.051	0.146	0.176	0.098	0.022	0.001		
	10			0.001	0.022	0.098	0.176	0.146	0.051	0.005		
	11				0.008	0.053	0.144	0.180	0.098	0.017		
	12				0.002	0.024	0.096	0.180	0.153	0.044	0.001	
	13				0.001	0.008	0.052	0.145	0.192	0.095	0.007	
	14					0.002	0.022	0.093	0.192	0.164	0.027	0.002
	15					0.001	0.007	0.047	0.149	0.218	0.080	0.011
	16						0.002	0.017	0.087	0.218	0.180	0.053
	17							0.005	0.036	0.154	0.285	0.179
	18							0.001	0.009	0.068	0.285	0.377
	19								0.001	0.014	0.135	0.377

Table B

The Binomial Distribution *Continued*

n	x	0.05	0.1	0.2	0.3	0.4	0.5	0.6	0.7	0.8	0.9	0.95
20	0	0.358	0.122	0.012	0.001							
	1	0.377	0.270	0.058	0.007							
	2	0.189	0.285	0.137	0.028	0.003						
	3	0.060	0.190	0.205	0.072	0.012	0.001					
	4	0.013	0.090	0.218	0.130	0.035	0.005					
	5	0.002	0.032	0.175	0.179	0.075	0.015	0.001				
	6		0.009	0.109	0.192	0.124	0.037	0.005				
	7		0.002	0.055	0.164	0.166	0.074	0.015	0.001			
	8			0.022	0.114	0.180	0.120	0.035	0.004			
	9			0.007	0.065	0.160	0.160	0.071	0.012			
	10			0.002	0.031	0.117	0.176	0.117	0.031	0.002		
	11				0.012	0.071	0.160	0.160	0.065	0.007		
	12				0.004	0.035	0.120	0.180	0.114	0.022		
	13				0.001	0.015	0.074	0.166	0.164	0.055	0.002	
	14					0.005	0.037	0.124	0.192	0.109	0.009	
	15					0.001	0.015	0.075	0.179	0.175	0.032	0.002
	16						0.005	0.035	0.130	0.218	0.090	0.013
	17						0.001	0.012	0.072	0.205	0.190	0.060
	18							0.003	0.028	0.137	0.285	0.189
	19								0.007	0.058	0.270	0.377
	20								0.001	0.012	0.122	0.358

Table C

The Poisson Distribution

x	0.1	0.2	0.3	0.4	0.5	0.6	0.7	0.8	0.9	1.0
0	.9048	.8187	.7408	.6703	.6065	.5488	.4966	.4493	.4066	.3679
1	.0905	.1637	.2222	.2681	.3033	.3293	.3476	.3595	.3659	.3679
2	.0045	.0164	.0333	.0536	.0758	.0988	.1217	.1438	.1647	.1839
3	.0002	.0011	.0033	.0072	.0126	.0198	.0284	.0383	.0494	.0613
4	.0000	.0001	.0003	.0007	.0016	.0030	.0050	.0077	.0111	.0153
5	.0000	.0000	.0000	.0001	.0002	.0004	.0007	.0012	.0020	.0031
6	.0000	.0000	.0000	.0000	.0000	.0000	.0001	.0002	.0003	.0005
7	.0000	.0000	.0000	.0000	.0000	.0000	.0000	.0000	.0000	.0001

Source: From Beyer, W. H., *Handbook of Tables for Probability and Statistics, 2nd Edition,* CRC Press, Boca Raton, Florida, 1986. With permission.

Table C

The Poisson Distribution *Continued*

					λ					
x	1.1	1.2	1.3	1.4	1.5	1.6	1.7	1.8	1.9	2.0
0	.3329	.3012	.2725	.2466	.2231	.2019	.1827	.1653	.1496	.1353
1	.3662	.3614	.3543	.3452	.3347	.3230	.3106	.2975	.2842	.2707
2	.2014	.2169	.2303	.2417	.2510	.2584	.2640	.2678	.2700	.2707
3	.0738	.0867	.0998	.1128	.1255	.1378	.1496	.1607	.1710	.1804
4	.0203	.0260	.0324	.0395	.0471	.0551	.0636	.0723	.0812	.0902
5	.0045	.0062	.0084	.0111	.0141	.0176	.0216	.0260	.0309	.0361
6	.0008	.0012	.0018	.0026	.0035	.0047	.0061	.0078	.0098	.0120
7	.0001	.0002	.0003	.0005	.0008	.0011	.0015	.0020	.0027	.0034
8	.0000	.0000	.0001	.0001	.0001	.0002	.0003	.0005	.0006	.0009
9	.0000	.0000	.0000	.0000	.0000	.0000	.0001	.0001	.0001	.0002

					λ					
x	2.1	2.2	2.3	2.4	2.5	2.6	2.7	2.8	2.9	3.0
0	.1225	.1108	.1003	.0907	.0821	.0743	.0672	.0608	.0550	.0498
1	.2572	.2438	.2306	.2177	.2052	.1931	.1815	.1703	.1596	.1494
2	.2700	.2681	.2652	.2613	.2565	.2510	.2450	.2384	.2314	.2240
3	.1890	.1966	.2033	.2090	.2138	.2176	.2205	.2225	.2237	.2240
4	.0992	.1082	.1169	.1254	.1336	.1414	.1488	.1557	.1622	.1680
5	.0417	.0476	.0538	.0602	.0668	.0735	.0804	.0872	.0940	.1008
6	.0146	.0174	.0206	.0241	.0278	.0319	.0362	.0407	.0455	.0504
7	.0044	.0055	.0068	.0083	.0099	.0118	.0139	.0163	.0188	.0216
8	.0011	.0015	.0019	.0025	.0031	.0038	.0047	.0057	.0068	.0081
9	.0003	.0004	.0005	.0007	.0009	.0011	.0014	.0018	.0022	.0027
10	.0001	.0001	.0001	.0002	.0002	.0003	.0004	.0005	.0006	.0008
11	.0000	.0000	.0000	.0000	.0000	.0001	.0001	.0001	.0002	.0002
12	.0000	.0000	.0000	.0000	.0000	.0000	.0000	.0000	.0000	.0001

					λ					
x	3.1	3.2	3.3	3.4	3.5	3.6	3.7	3.8	3.9	4.0
0	.0450	.0408	.0369	.0334	.0302	.0273	.0247	.0224	.0202	.0183
1	.1397	.1304	.1217	.1135	.1057	.0984	.0915	.0850	.0789	.0733
2	.2165	.2087	.2008	.1929	.1850	.1771	.1692	.1615	.1539	.1465
3	.2237	.2226	.2209	.2186	.2158	.2125	.2087	.2046	.2001	.1954
4	.1734	.1781	.1823	.1858	.1888	.1912	.1931	.1944	.1951	.1954

Table C

The Poisson Distribution *Continued*

	λ									
x	3.1	3.2	3.3	3.4	3.5	3.6	3.7	3.8	3.9	4.0
5	.1075	.1140	.1203	.1264	.1322	.1377	.1429	.1477	.1522	.1563
6	.0555	.0608	.0662	.0716	.0771	.0826	.0881	.0936	.0989	.1042
7	.0246	.0278	.0312	.0348	.0385	.0425	.0466	.0508	.0551	.0595
8	.0095	.0111	.0129	.0148	.0169	.0191	.0215	.0241	.0269	.0298
9	.0033	.0040	.0047	.0056	.0066	.0076	.0089	.0102	.0116	.0132
10	.0010	.0013	.0016	.0019	.0023	.0028	.0033	.0039	.0045	.0053
11	.0003	.0004	.0005	.0006	.0007	.0009	.0011	.0013	.0016	.0019
12	.0001	.0001	.0001	.0002	.0002	.0003	.0003	.0004	.0005	.0006
13	.0000	.0000	.0000	.0000	.0001	.0001	.0001	.0001	.0002	.0002
14	.0000	.0000	.0000	.0000	.0000	.0000	.0000	.0000	.0000	.0001

	λ									
x	4.1	4.2	4.3	4.4	4.5	4.6	4.7	4.8	4.9	5.0
0	.0166	.0150	.0136	.0123	.0111	.0101	.0091	.0082	.0074	.0067
1	.0679	.0630	.0583	.0540	.0500	.0462	.0427	.0395	.0365	.0337
2	.1393	.1323	.1254	.1188	.1125	.1063	.1005	.0948	.0894	.0842
3	.1904	.1852	.1798	.1743	.1687	.1631	.1574	.1517	.1460	.1404
4	.1951	.1944	.1933	.1917	.1898	.1875	.1849	.1820	.1789	.1755
5	.1600	.1633	.1662	.1687	.1708	.1725	.1738	.1747	.1753	.1755
6	.1093	.1143	.1191	.1237	.1281	.1323	.1362	.1398	.1432	.1462
7	.0640	.0686	.0732	.0778	.0824	.0869	.0914	.0959	.1002	.1044
8	.0328	.0360	.0393	.0428	.0463	.0500	.0537	.0575	.0614	.0653
9	.0150	.0168	.0188	.0209	.0232	.0255	.0280	.0307	.0334	.0363
10	.0061	.0071	.0081	.0092	.0104	.0118	.0132	.0147	.0164	.0181
11	.0023	.0027	.0032	.0037	.0043	.0049	.0056	.0064	.0073	.0082
12	.0008	.0009	.0011	.0014	.0016	.0019	.0022	.0026	.0030	.0034
13	.0002	.0003	.0004	.0005	.0006	.0007	.0008	.0009	.0011	.0013
14	.0001	.0001	.0001	.0001	.0002	.0002	.0003	.0003	.0004	.0005
15	.0000	.0000	.0000	.0000	.0001	.0001	.0001	.0001	.0001	.0002

	λ									
x	5.1	5.2	5.3	5.4	5.5	5.6	5.7	5.8	5.9	6.0
0	.0061	.0055	.0050	.0045	.0041	.0037	.0033	.0030	.0027	.0025
1	.0311	.0287	.0265	.0244	.0225	.0207	.0191	.0176	.0162	.0149
2	.0793	.0746	.0701	.0659	.0618	.0580	.0544	.0509	.0477	.0446
3	.1348	.1293	.1239	.1185	.1133	.1082	.1033	.0985	.0938	.0892
4	.1719	.1681	.1641	.1600	.1558	.1515	.1472	.1428	.1383	.1339

Table C

The Poisson Distribution *Continued*

					λ					
x	5.1	5.2	5.3	5.4	5.5	5.6	5.7	5.8	5.9	6.0
5	.1753	.1748	.1740	.1728	.1714	.1697	.1678	.1656	.1632	.1606
6	.1490	.1515	.1537	.1555	.1571	.1584	.1594	.1601	.1605	.1606
7	.1086	.1125	.1163	.1200	.1234	.1267	:1298	.1326	.1353	.1377
8	.0692	.0731	.0771	.0810	.0849	.0887	.0925	.0962	.0998	.1033
9	.0392	.0423	.0454	.0486	.0519	.0552	.0586	.0620	.0654	.0688
10	.0200	.0220	.0241	.0262	.0285	.0309	.0334	.0359	.0386	.0413
11	.0093	.0104	.0116	.0129	.0143	.0157	.0173	.0190	.0207	.0225
12	.0039	.0045	.0051	.0058	.0065	.0073	.0082	.0092	.0102	.0113
13	.0015	.0018	.0021	.0024	.0028	.0032	.0036	.0041	.0046	.0052
14	.0006	.0007	.0008	.0009	.0011	.0013	.0015	.0017	.0019	.0022
15	.0002	.0002	.0003	.0003	.0004	.0005	.0006	.0007	.0008	.0009
16	.0001	.0001	.0001	.0001	.0001	.0002	.0002	.0002	.0003	.0003
17	.0000	.0000	.0000	.0000	.0000	.0000	.0001	.0001	.0001	.0001

					λ					
x	6.1	6.2	6.3	6.4	6.5	6.6	6.7	6.8	6.9	7.0
0	.0022	.0020	.0018	.0017	.0015	.0014	.0012	.0011	.0010	.0009
1	.0137	.0126	.0116	.0106	.0098	.0090	.0082	.0076	.0070	.0064
2	.0417	.0390	.0364	.0340	.0318	.0296	.0276	.0258	.0240	.0223
3	.0848	.0806	.0765	.0726	.0688	.0652	.0617	.0584	.0552	.0521
4	.1294	.1249	.1205	.1162	.1118	.1076	.1034	.0992	.0952	.0912
5	.1579	.1549	.1519	.1487	.1454	.1420	.1385	.1349	.1314	.1277
6	.1605	.1601	.1595	.1586	.1575	.1562	.1546	.1529	.1511	.1490
7	.1399	.1418	.1435	.1450	.1462	.1472	.1480	.1486	.1489	.1490
8	.1066	.1099	.1130	.1160	.1188	.1215	.1240	.1263	.1284	.1304
9	.0723	.0757	.0791	.0825	.0858	.0891	.0923	.0954	.0985	.1014
10	.0441	.0469	.0498	.0528	.0558	.0588	.0618	.0649	.0679	.0710
11	.0245	.0265	.0285	.0307	.0330	.0353	.0377	.0401	.0426	.0452
12	.0124	.0137	.0150	.0164	.0179	.0194	.0210	.0227	.0245	.0264
13	.0058	.0065	.0073	.0081	.0089	.0098	.0108	.0119	.0130	.0142
14	.0025	.0029	.0033	.0037	.0041	.0046	.0052	.0058	.0064	.0071
15	.0010	.0012	.0014	.0016	.0018	.0020	.0023	.0026	.0029	.0033
16	.0004	.0005	.0005	.0006	.0007	.0008	.0010	.0011	.0013	.0014
17	.0001	.0002	.0002	.0002	.0003	.0003	.0004	.0004	.0005	.0006
18	.0000	.0001	.0001	.0001	.0001	.0001	.0001	.0002	.0002	.0002
19	.0000	.0000	.0000	.0000	.0000	.0000	.0000	.0001	.0001	.0001

Table C

The Poisson Distribution   *Continued*

					λ					
x	7.1	7.2	7.3	7.4	7.5	7.6	7.7	7.8	7.9	8.0
0	.0008	.0007	.0007	.0006	.0006	.0005	.0005	.0004	.0004	.0003
1	.0059	.0054	.0049	.0045	.0041	.0038	.0035	.0032	.0029	.0027
2	.0208	.0194	.0180	.0167	.0156	.0145	.0134	.0125	.0116	.0107
3	.0492	.0464	.0438	.0413	.0389	.0366	.0345	.0324	.0305	.0286
4	.0874	.0836	.0799	.0764	.0729	.0696	.0663	.0632	.0602	.0573
5	.1241	.1204	.1167	.1130	.1094	.1057	.1021	.0986	.0951	.0916
6	.1468	.1445	.1420	.1394	.1367	.1339	.1311	.1282	.1252	.1221
7	.1489	.1486	.1481	.1474	.1465	.1454	.1442	.1428	.1413	.1396
8	.1321	.1337	.1351	.1363	.1373	.1382	.1388	.1392	.1395	.1396
9	.1042	.1070	.1096	.1121	.1144	.1167	.1187	.1207	.1224	.1241
10	.0740	.0770	.0800	.0829	.0858	.0887	.0914	.0941	.0967	.0993
11	.0478	.0504	.0531	.0558	.0585	.0613	.0640	.0667	.0695	.0722
12	.0283	.0303	.0323	.0344	.0366	.0388	.0411	.0434	.0457	.0481
13	.0154	.0168	.0181	.0196	.0211	.0227	.0243	.0260	.0278	.0296
14	.0078	.0086	.0095	.0104	.0113	.0123	.0134	.0145	.0157	.0169
15	.0037	.0041	.0046	.0051	.0057	.0062	.0069	.0075	.0083	.0090
16	.0016	.0019	.0021	.0024	.0026	.0030	.0033	.0037	.0041	.0045
17	.0007	.0008	.0009	.0010	.0012	.0013	.0015	.0017	.0019	.0021
18	.0003	.0003	.0004	.0004	.0005	.0006	.0006	.0007	.0008	.0009
19	.0001	.0001	.0001	.0002	.0002	.0002	.0003	.0003	.0003	.0004
20	.0000	.0000	.0001	.0001	.0001	.0001	.0001	.0001	.0001	.0002
21	.0000	.0000	.0000	.0000	.0000	.0000	.0000	.0000	.0001	.0001

					λ					
x	8.1	8.2	8.3	8.4	8.5	8.6	8.7	8.8	8.9	9.0
0	.0003	.0003	.0002	.0002	.0002	.0002	.0002	.0002	.0001	.0001
1	.0025	.0023	.0021	.0019	.0017	.0016	.0014	.0013	.0012	.0011
2	.0100	.0092	.0086	.0079	.0074	.0068	.0063	.0058	.0054	.0050
3	.0269	.0252	.0237	.0222	.0208	.0195	.0183	.0171	.0160	.0150
4	.0544	.0517	.0491	.0466	.0443	.0420	.0398	.0377	.0357	.0337
5	.0882	.0849	.0816	.0784	.0752	.0722	.0692	.0663	.0635	.0607
6	.1191	.1160	.1128	.1097	.1066	.1034	.1003	.0972	.0941	.0911
7	.1378	.1358	.1338	.1317	.1294	.1271	.1247	.1222	.1197	.1171
8	.1395	.1392	.1388	.1382	.1375	.1366	.1356	.1344	.1332	.1318
9	.1256	.1269	.1280	.1290	.1299	.1306	.1311	.1315	.1317	.1318

Table C

The Poisson Distribution *Continued*

x	8.1	8.2	8.3	8.4	8.5	8.6	8.7	8.8	8.9	9.0
10	.1017	.1040	.1063	.1084	.1104	.1123	.1140	.1157	.1172	.1186
11	.0749	.0776	.0802	.0828	.0853	.0878	.0902	.0925	.0948	.0970
12	.0505	.0530	.0555	.0579	.0604	.0629	.0654	.0679	.0703	.0728
13	.0315	.0334	.0354	.0374	.0395	.0416	.0438	.0459	.0481	.0504
14	.0182	.0196	.0210	.0225	.0240	.0256	.0272	.0289	.0306	.0324
15	.0098	.0107	.0116	.0126	.0136	.0147	.0158	.0169	.0182	.0194
16	.0050	.0055	.0060	.0066	.0072	.0079	.0086	.0093	.0101	.0109
17	.0024	.0026	.0029	.0033	.0036	.0040	.0044	.0048	.0053	.0058
18	.0011	.0012	.0014	.0015	.0017	.0019	.0021	.0024	.0026	.0029
19	.0005	.0005	.0006	.0007	.0008	.0009	.0010	.0011	.0012	.0014
20	.0002	.0002	.0002	.0003	.0003	.0004	.0004	.0005	.0005	.0006
21	.0001	.0001	.0001	.0001	.0001	.0002	.0002	.0002	.0002	.0003
22	.0000	.0000	.0000	.0000	.0001	.0001	.0001	.0001	.0001	.0001

x	9.1	9.2	9.3	9.4	9.5	9.6	9.7	9.8	9.9	10
0	.0001	.0001	.0001	.0001	.0001	.0001	.0001	.0001	.0001	.0000
1	.0010	.0009	.0009	.0008	.0007	.0007	.0006	.0005	.0005	.0005
2	.0046	.0043	.0040	.0037	.0034	.0031	.0029	.0027	.0025	.0023
3	.0140	.0131	.0123	.0115	.0107	.0100	.0093	.0087	.0081	.0076
4	.0319	.0302	.0285	.0269	.0254	.0240	.0226	.0213	.0201	.0189
5	.0581	.0555	.0530	.0506	.0483	.0460	.0439	.0418	.0398	.0378
6	.0881	.0851	.0822	.0793	.0764	.0736	.0709	.0682	.0656	.0631
7	.1145	.1118	.1091	.1064	.1037	.1010	.0982	.0955	.0928	.0901
8	.1302	.1286	.1269	.1251	.1232	.1212	.1191	.1170	.1148	.1126
9	.1317	.1315	.1311	.1306	.1300	.1293	.1284	.1274	.1263	.1251
10	.1198	.1210	.1219	.1228	.1235	.1241	.1245	.1249	.1250	.1251
11	.0991	.1012	.1031	.1049	.1067	.1083	.1098	.1112	.1125	.1137
12	.0752	.0776	.0799	.0822	.0844	.0866	.0888	.0908	.0928	.0948
13	.0526	.0549	.0572	.0594	.0617	.0640	.0662	.0685	.0707	.0729
14	.0342	.0361	.0380	.0399	.0419	.0439	.0459	.0479	.0500	.0521
15	.0208	.0221	.0235	.0250	.0265	.0281	.0297	.0313	.0330	.0347
16	.0118	.0127	.0137	.0147	.0157	.0168	.0180	.0192	.0204	.0217
17	.0063	.0069	.0075	.0081	.0088	.0095	.0103	.0111	.0119	.0128
18	.0032	.0035	.0039	.0042	.0046	.0051	.0055	.0060	.0065	.0071
19	.0015	.0017	.0019	.0021	.0023	.0026	.0028	.0031	.0034	.0037

Table C

The Poisson Distribution   *Continued*

x	9.1	9.2	9.3	9.4	9.5	9.6	9.7	9.8	9.9	10
20	.0007	.0008	.0009	.0010	.0011	.0012	.0014	.0015	.0017	.0019
21	.0003	.0003	.0004	.0004	.0005	.0006	.0006	.0007	.0008	.0009
22	.0001	.0001	.0002	.0002	.0002	.0002	.0003	.0003	.0004	.0004
23	.0000	.0001	.0001	.0001	.0001	.0001	.0001	.0001	.0002	.0002
24	.0000	.0000	.0000	.0000	.0000	.0000	.0000	.0001	.0001	.0001

λ

x	11	12	13	14	15	16	17	18	19	20
0	.0000	.0000	.0000	.0000	.0000	.0000	.0000	.0000	.0000	.0000
1	.0002	.0001	.0000	.0000	.0000	.0000	.0000	.0000	.0000	.0000
2	.0010	.0004	.0002	.0001	.0000	.0000	.0000	.0000	.0000	.0000
3	.0037	.0018	.0008	.0004	.0002	.0001	.0000	.0000	.0000	.0000
4	.0102	.0053	.0027	.0013	.0006	.0003	.0001	.0001	.0000	.0000
5	.0224	.0127	.0070	.0037	.0019	.0010	.0005	.0002	.0001	.0001
6	.0411	.0255	.0152	.0087	.0048	.0026	.0014	.0007	.0004	.0002
7	.0646	.0437	.0281	.0174	.0104	.0060	.0034	.0018	.0010	.0005
8	.0888	.0655	.0457	.0304	.0194	.0120	.0072	.0042	.0024	.0013
9	.1085	.0874	.0661	.0473	.0324	.0213	.0135	.0083	.0050	.0029
10	.1194	.1048	.0859	.0663	.0486	.0341	.0230	.0150	.0095	.0058
11	.1194	.1144	.1015	.0844	.0663	.0496	.0355	.0245	.0164	.0106
12	.1094	.1144	.1099	.0984	.0829	.0661	.0504	.0368	.0259	.0176
13	.0926	.1056	.1099	.1060	.0956	.0814	.0658	.0509	.0378	.0271
14	.0728	.0905	.1021	.1060	.1024	.0930	.0800	.0655	.0514	.0387
15	.0534	.0724	.0885	.0989	.1024	.0992	.0906	.0786	.0650	.0516
16	.0367	.0543	.0719	.0866	.0960	.0992	.0963	.0884	.0772	.0646
17	.0237	.0383	.0550	.0713	.0847	.0934	.0963	.0936	.0863	.0760
18	.0145	.0256	.0397	.0554	.0706	.0830	.0909	.0936	.0911	.0844
19	.0084	.0161	.0272	.0409	.0557	.0699	.0814	.0887	.0911	.0888
20	.0046	.0097	.0177	.0286	.0418	.0559	.0692	.0798	.0866	.0888
21	.0024	.0055	.0109	.0191	.0299	.0426	.0560	.0684	.0783	.0846
22	.0012	.0030	.0065	.0121	.0204	.0310	.0433	.0560	.0676	.0769
23	.0006	.0016	.0037	.0074	.0133	.0216	.0320	.0438	.0559	.0669
24	.0003	.0008	.0020	.0043	.0083	.0144	.0226	.0328	.0442	.0557
25	.0001	.0004	.0010	.0024	.0050	.0092	.0154	.0237	.0336	.0446
26	.0000	.0002	.0005	.0013	.0029	.0057	.0101	.0164	.0246	.0343
27	.0000	.0001	.0002	.0007	.0016	.0034	.0063	.0109	.0173	.0254
28	.0000	.0000	.0001	.0003	.0009	.0019	.0038	.0070	.0117	.0181
29	.0000	.0000	.0001	.0002	.0004	.0011	.0023	.0044	.0077	.0125

Table C

The Poisson Distribution   *Continued*

x	11	12	13	14	15	16	17	18	19	20
30	.0000	.0000	.0000	.0001	.0002	.0006	.0013	.0026	.0049	.0083
31	.0000	.0000	.0000	.0000	.0001	.0003	.0007	.0015	.0030	.0054
32	.0000	.0000	.0000	.0000	.0001	.0001	.0004	.0009	.0018	.0034
33	.0000	.0000	.0000	.0000	.0000	.0001	.0002	.0005	.0010	.0020
34	.0000	.0000	.0000	.0000	.0000	.0000	.0001	.0002	.0006	.0012
35	.0000	.0000	.0000	.0000	.0000	.0000	.0000	.0001	.0003	.0007
36	.0000	.0000	.0000	.0000	.0000	.0000	.0000	.0001	.0002	.0004
37	.0000	.0000	.0000	.0000	.0000	.0000	.0000	.0000	.0001	.0002
38	.0000	.0000	.0000	.0000	.0000	.0000	.0000	.0000	.0000	.0001
39	.0000	.0000	.0000	.0000	.0000	.0000	.0000	.0000	.0000	.0001

Table D

Random Numbers

10480	15011	01536	02011	81647	91646	69179	14194	62590	36207	20969	99570	91291	90700
22368	46573	25595	85393	30995	89198	27982	53402	93965	34095	52666	19174	39615	99505
24130	48360	22527	97265	76393	64809	15179	24830	49340	32081	30680	19655	63348	58629
42167	93093	06243	61680	07856	16376	39440	53537	71341	57004	00849	74917	97758	16379
37570	39975	81837	16656	06121	91782	60468	81305	49684	60672	14110	06927	01263	54613
77921	06907	11008	42751	27756	53498	18602	70659	90655	15053	21916	81825	44394	42880
99562	72905	56420	69994	98872	31016	71194	18738	44013	48840	63213	21069	10634	12952
96301	91977	05463	07972	18876	20922	94595	56869	69014	60045	18425	84903	42508	32307
89579	14342	63661	10281	17453	18103	57740	84378	25331	12566	58678	44947	05584	56941
85475	36857	43342	53988	53060	59533	38867	62300	08158	17983	16439	11458	18593	64952
28918	69578	88231	33276	70997	79936	56865	05859	90106	31595	01547	85590	91610	78188
63553	40961	48235	03427	49626	69445	18663	72695	52180	20847	12234	90511	33703	90322
09429	93969	52636	92737	88974	33488	36320	17617	30015	08272	84115	27156	30613	74952
10365	61129	87529	85689	48237	52267	67689	93394	01511	26358	85104	20285	29975	89868
07119	97336	71048	08178	77233	13916	47564	81056	97735	85977	29372	74461	28551	90707
51085	12765	51821	51259	77452	16308	60756	92144	49442	53900	70960	63990	75601	40719
02368	21382	52404	60268	89368	19885	55322	44819	01188	65255	64835	44919	05944	55157
01011	54092	33362	94904	31273	04146	18594	29852	71585	85030	51132	01915	92747	64951
52162	53916	46369	58586	23216	14513	83149	98736	23495	64350	94738	17752	35156	35749
07056	97628	33787	09998	42698	06691	76988	13602	51851	46104	88916	19509	25625	58104

Table D

Random Numbers *Continued*

48663	91245	85828	14346	09172	30168	90229	04734	59193	22178	30421	61666	99904	32812
54164	58492	22421	74103	47070	25306	76468	26384	58151	06646	21524	15227	96909	44592
32639	32363	05597	24200	13363	38005	94342	28728	35806	06912	17012	64161	18296	22851
29334	27001	87637	87308	58731	00256	45834	15398	46557	41135	10367	07684	36188	18510
02488	33062	28834	07351	19731	92420	60952	61280	50001	67658	32586	86679	50720	94953
81525	72295	04839	96423	24878	82651	66566	14778	76797	14780	13300	87074	79666	95725
29676	20591	68086	26432	46901	20849	89768	81536	86645	12659	92259	57102	80428	25280
00742	57392	39064	66432	84673	40027	32832	61362	98947	96067	64760	64584	96096	98253
05366	04213	25669	26422	44407	44048	37937	63904	45766	66134	75470	66520	34693	90449
91921	26418	64117	94305	26766	25940	39972	22209	71500	64568	91402	42416	07844	69618
00582	04711	87917	77341	42206	35126	74087	99547	81817	42607	43808	76655	62028	76630
00725	69884	62797	56170	86324	88072	76222	36086	84637	93161	76038	65855	77919	88006
69011	65797	95876	55293	18988	27354	26575	08625	40801	59920	29841	80150	12777	48501
25976	57948	29888	88604	67917	48708	18912	82271	65424	69774	33611	54262	85963	03547
09763	83473	73577	12908	30883	18317	28290	35797	05998	41688	34952	37888	38917	88050
91567	42595	27958	30134	04024	86385	29880	99730	55536	84855	29080	09250	79656	73211
17955	56349	90999	49127	20044	59931	06115	20542	18059	02008	73708	83517	36103	42791
46503	18584	18845	49618	02304	51038	20655	58727	28168	15475	56942	53389	20562	87338
92157	89634	94824	78171	84610	82834	09922	25417	44137	48413	25555	21246	35509	20468
14577	62765	35605	81263	39667	47358	56873	56307	61607	49518	89656	20103	77490	18062
98427	07523	33362	64270	01638	92477	66969	98420	04880	45585	46565	04102	46880	45709
34914	63976	88720	82765	34476	17032	87589	40836	32427	70002	70663	88863	77775	69348
70060	28277	39475	46473	23219	53416	94970	25832	69975	94884	19661	72828	00102	66794
53976	54914	06990	67245	68350	82948	11398	42878	80287	88267	47363	46634	06541	97809
76072	29515	40980	07391	58745	25774	22987	80059	39911	96189	41151	14222	60697	59583
90725	52210	83974	29992	65831	38857	50490	83765	55657	14361	31720	57375	56228	41546
64364	67412	33339	31926	14883	24413	59744	92351	97473	89286	35931	04110	23726	51900
08962	00358	31662	25388	61642	34072	81249	35648	56891	69352	48373	45578	78547	81788
95012	68379	93526	70765	10593	04542	76463	54328	02349	17247	28865	14777	62730	92277
15664	10493	20492	38391	91132	21999	59516	81652	27195	48223	46751	22923	32261	85653

Source: From Beyer, W. H., *Handbook of Tables for Probability and Statistics, 2nd Edition,* CRC Press, Boca Raton, Florida, 1986. With permission.

Table E

The Standard Normal Distribution

z	.00	.01	.02	.03	.04	.05	.06	.07	.08	.09
0.0	.0000	.0040	.0080	.0120	.0160	.0199	.0239	.0279	.0319	.0359
0.1	.0398	.0438	.0478	.0517	.0557	.0596	.0636	.0675	.0714	.0753
0.2	.0793	.0832	.0871	.0910	.0948	.0987	.1026	.1064	.1103	.1141
0.3	.1179	.1217	.1255	.1293	.1331	.1368	.1406	.1443	.1480	.1517
0.4	.1554	.1591	.1628	.1664	.1700	.1736	.1772	.1808	.1844	.1879
0.5	.1915	.1950	.1985	.2019	.2054	.2088	.2123	.2157	.2190	.2224
0.6	.2257	.2291	.2324	.2357	.2389	.2422	.2454	.2486	.2517	.2549
0.7	.2580	.2611	.2642	.2673	.2704	.2734	.2764	.2794	.2823	.2852
0.8	.2881	.2910	.2939	.2967	.2995	.3023	.3051	.3078	.3106	.3133
0.9	.3159	.3186	.3212	.3238	.3264	.3289	.3315	.3340	.3365	.3389
1.0	.3413	.3438	.3461	.3485	.3508	.3531	.3554	.3577	.3599	.3621
1.1	.3643	.3665	.3686	.3708	.3729	.3749	.3770	.3790	.3810	.3830
1.2	.3849	.3869	.3888	.3907	.3925	.3944	.3962	.3980	.3997	.4015
1.3	.4032	.4049	.4066	.4082	.4099	.4115	.4131	.4147	.4162	.4177
1.4	.4192	.4207	.4222	.4236	.4251	.4265	.4279	.4292	.4306	.4319
1.5	.4332	.4345	.4357	.4370	.4382	.4394	.4406	.4418	.4429	.4441
1.6	.4452	.4463	.4474	.4484	.4495	.4505	.4515	.4525	.4535	.4545
1.7	.4554	.4564	.4573	.4582	.4591	.4599	.4608	.4616	.4625	.4633
1.8	.4641	.4649	.4656	.4664	.4671	.4678	.4686	.4693	.4699	.4706
1.9	.4713	.4719	.4726	.4732	.4738	.4744	.4750	.4756	.4761	.4767
2.0	.4772	.4778	.4783	.4788	.4793	.4798	.4803	.4808	.4812	.4817
2.1	.4821	.4826	.4830	.4834	.4838	.4842	.4846	.4850	.4854	.4857
2.2	.4861	.4864	.4868	.4871	.4875	.4878	.4881	.4884	.4887	.4890
2.3	.4893	.4896	.4898	.4901	.4904	.4906	.4909	.4911	.4913	.4916
2.4	.4918	.4920	.4922	.4925	.4927	.4929	.4931	.4932	.4934	.4936
2.5	.4938	.4940	.4941	.4943	.4945	.4946	.4948	.4949	.4951	.4952
2.6	.4953	.4955	.4956	.4957	.4959	.4960	.4961	.4962	.4963	.4964
2.7	.4965	.4966	.4967	.4968	.4969	.4970	.4971	.4972	.4973	.4974
2.8	.4974	.4975	.4976	.4977	.4977	.4978	.4979	.4979	.4980	.4981
2.9	.4981	.4982	.4982	.4983	.4984	.4984	.4985	.4985	.4986	.4986
3.0	.4987	.4987	.4987	.4988	.4988	.4989	.4989	.4989	.4990	.4990

Source: Frederick Mosteller and Robert E. K. Rourke, *Sturdy Statistics,* Table A–1 (Reading, Mass.: Addison-Wesley, 1973). Reprinted with permission of the copyright owners.

Note: Use 0.4999 for z values above 3.09.

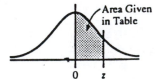

Table F

The *t* Distribution

	Confidence Intervals	50%	80%	90%	95%	98%	99%
	ONE TAIL, α	0.25	0.10	0.05	0.025	0.01	0.005
d.f.	TWO TAILS, α	0.50	0.20	0.10	0.05	0.02	0.01
1		1.000	3.078	6.314	12.706	31.821	63.657
2		.816	1.886	2.920	4.303	6.965	9.925
3		.765	1.638	2.353	3.182	4.541	5.841
4		.741	1.533	2.132	2.776	3.747	4.604
5		.727	1.476	2.015	2.571	3.365	4.032
6		.718	1.440	1.943	2.447	3.143	3.707
7		.711	1.415	1.895	2.365	2.998	3.499
8		.706	1.397	1.860	2.306	2.896	3.355
9		.703	1.383	1.833	2.262	2.821	3.250
10		.700	1.372	1.812	2.228	2.764	3.169
11		.697	1.363	1.796	2.201	2.718	3.106
12		.695	1.356	1.782	2.179	2.681	3.055
13		.694	1.350	1.771	2.160	2.650	3.012
14		.692	1.345	1.761	2.145	2.624	2.977
15		.691	1.341	1.753	2.131	2.602	2.947
16		.690	1.337	1.746	2.120	2.583	2.921
17		.689	1.333	1.740	2.110	2.567	2.898
18		.688	1.330	1.734	2.101	2.552	2.878
19		.688	1.328	1.729	2.093	2.539	2.861
20		.687	1.325	1.725	2.086	2.528	2.845
21		.686	1.323	1.721	2.080	2.518	2.831
22		.686	1.321	1.717	2.074	2.508	2.819
23		.685	1.319	1.714	2.069	2.500	2.807
24		.685	1.318	1.711	2.064	2.492	2.797
25		.684	1.316	1.708	2.060	2.485	2.787
26		.684	1.315	1.706	2.056	2.479	2.779
27		.684	1.314	1.703	2.052	2.473	2.771
28		.683	1.313	1.701	2.048	2.467	2.763
(z) ∞		.674	1.282[a]	1.645[b]	1.960	2.326[c]	2.576[d]

Source: Adapted from Beyer, W. H., *Handbook of Tables for Probability and Statistics, 2nd Edition,* CRC Press, Boca Raton, Florida, 1986. With permission.

[a]This value has been rounded to 1.28 in the textbook.
[b]This value has been rounded to 1.65 in the textbook.
[c]This value has been rounded to 2.33 in the textbook.
[d]This value has been rounded to 2.58 in the textbook.

One Tail

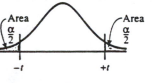

Area α

Two Tails

Area α/2   Area α/2

−t   +t

Table G

The Chi-Square Distribution

Degrees of freedom	α									
	0.995	0.99	0.975	0.95	0.90	0.10	0.05	0.025	0.01	0.005
1	—	—	0.001	0.004	0.016	2.706	3.841	5.024	6.635	7.879
2	0.010	0.020	0.051	0.103	0.211	4.605	5.991	7.378	9.210	10.597
3	0.072	0.115	0.216	0.352	0.584	6.251	7.815	9.348	11.345	12.838
4	0.207	0.297	0.484	0.711	1.064	7.779	9.488	11.143	13.277	14.860
5	0.412	0.554	0.831	1.145	1.610	9.236	11.071	12.833	15.086	16.750
6	0.676	0.872	1.237	1.635	2.204	10.645	12.592	14.449	16.812	18.548
7	0.989	1.239	1.690	2.167	2.833	12.017	14.067	16.013	18.475	20.278
8	1.344	1.646	2.180	2.733	3.490	13.362	15.507	17.535	20.090	21.955
9	1.735	2.088	2.700	3.325	4.168	14.684	16.919	19.023	21.666	23.589
10	2.156	2.558	3.247	3.940	4.865	15.987	18.307	20.483	23.209	25.188
11	2.603	3.053	3.816	4.575	5.578	17.275	19.675	21.920	24.725	26.757
12	3.074	3.571	4.404	5.226	6.304	18.549	21.026	23.337	26.217	28.299
13	3.565	4.107	5.009	5.892	7.042	19.812	22.362	24.736	27.688	29.819
14	4.075	4.660	5.629	6.571	7.790	21.064	23.685	26.119	29.141	31.319
15	4.601	5.229	6.262	7.261	8.547	22.307	24.996	27.488	30.578	32.801
16	5.142	5.812	6.908	7.962	9.312	23.542	26.296	28.845	32.000	34.267
17	5.697	6.408	7.564	8.672	10.085	24.769	27.587	30.191	33.409	35.718
18	6.265	7.015	8.231	9.390	10.865	25.989	28.869	31.526	34.805	37.156
19	6.844	7.633	8.907	10.117	11.651	27.204	30.144	32.852	36.191	38.582
20	7.434	8.260	9.591	10.851	12.443	28.412	31.410	34.170	37.566	39.997
21	8.034	8.897	10.283	11.591	13.240	29.615	32.671	35.479	38.932	41.401
22	8.643	9.542	10.982	12.338	14.042	30.813	33.924	36.781	40.289	42.796
23	9.262	10.196	11.689	13.091	14.848	32.007	35.172	38.076	41.638	44.181
24	9.886	10.856	12.401	13.848	15.659	33.196	36.415	39.364	42.980	45.559
25	10.520	11.524	13.120	14.611	16.473	34.382	37.652	40.646	44.314	46.928

Table G

The Chi-Square Distribution *Continued*

Degrees of freedom	α									
	0.995	0.99	0.975	0.95	0.90	0.10	0.05	0.025	0.01	0.005
26	11.160	12.198	13.844	15.379	17.292	35.563	38.885	41.923	45.642	48.290
27	11.808	12.879	14.573	16.151	18.114	36.741	40.113	43.194	46.963	49.645
28	12.461	13.565	15.308	16.928	18.939	37.916	41.337	44.461	48.278	50.993
29	13.121	14.257	16.047	17.708	19.768	39.087	42.557	45.722	49.588	52.336
30	13.787	14.954	16.791	18.493	20.599	40.256	43.773	46.979	50.892	53.672
40	20.707	22.164	24.433	26.509	29.051	51.805	55.758	59.342	63.691	66.766
50	27.991	29.707	32.357	34.764	37.689	63.167	67.505	71.420	76.154	79.490
60	35.534	37.485	40.482	43.188	46.459	74.397	79.082	83.298	88.379	91.952
70	43.275	45.442	48.758	51.739	55.329	85.527	90.531	95.023	100.425	104.215
80	51.172	53.540	57.153	60.391	64.278	96.578	101.879	106.629	112.329	116.321
90	59.196	61.754	65.647	69.126	73.291	107.565	113.145	118.136	124.116	128.299
100	67.328	70.065	74.222	77.929	82.358	118.498	124.342	129.561	135.807	140.169

Source: Donald B. Owen, *Handbook of Statistics Tables.* © 1962, by Addison-Wesley Publishing Co., Inc., Reading, Massachusetts. Table A–5. Reprinted with permission of the publisher.

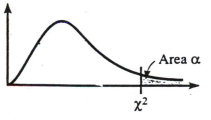

Table H

The F Distribution

d.f.D.: degrees of freedom, denominator	α = 0.005								
	d.f.N.: Degrees of freedom, numerator								
	1	*2*	*3*	*4*	*5*	*6*	*7*	*8*	*9*
1	16211	20000	21615	22500	23056	23437	23715	23925	24091
2	198.5	199.0	199.2	199.2	199.3	199.3	199.4	199.4	199.4
3	55.55	49.80	47.47	46.19	45.39	44.84	44.43	44.13	43.88
4	31.33	26.28	24.26	23.15	22.46	21.97	21.62	21.35	21.14
5	22.78	18.31	16.53	15.56	14.94	14.51	14.20	13.96	13.77
6	18.63	14.54	12.92	12.03	11.46	11.07	10.79	10.57	10.39
7	16.24	12.40	10.88	10.05	9.52	9.16	8.89	8.68	8.51
8	14.69	11.04	9.60	8.81	8.30	7.95	7.69	7.50	7.34
9	13.61	10.11	8.72	7.96	7.47	7.13	6.88	6.69	6.54
10	12.83	9.43	8.08	7.34	6.87	6.54	6.30	6.12	5.97
11	12.23	8.91	7.60	6.88	6.42	6.10	5.86	5.68	5.54
12	11.75	8.51	7.23	6.52	6.07	5.76	5.52	5.35	5.20
13	11.37	8.19	6.93	6.23	5.79	5.48	5.25	5.08	4.94
14	11.06	7.92	6.68	6.00	5.56	5.26	5.03	4.86	4.72
15	10.80	7.70	6.48	5.80	5.37	5.07	4.85	4.67	4.54
16	10.58	7.51	6.30	5.64	5.21	4.91	4.69	4.52	4.38
17	10.38	7.35	6.16	5.50	5.07	4.78	4.56	4.39	4.25
18	10.22	7.21	6.03	5.37	4.96	4.66	4.44	4.28	4.14
19	10.07	7.09	5.92	5.27	4.85	4.56	4.34	4.18	4.04
20	9.94	6.99	5.82	5.17	4.76	4.47	4.26	4.09	3.96
21	9.83	6.89	5.73	5.09	4.68	4.39	4.18	4.01	3.88
22	9.73	6.81	5.65	5.02	4.61	4.32	4.11	3.94	3.81
23	9.63	6.73	5.58	4.95	4.54	4.26	4.05	3.88	3.75
24	9.55	6.66	5.52	4.89	4.49	4.20	3.99	3.83	3.69
25	9.48	6.60	5.46	4.84	4.43	4.15	3.94	3.78	3.64
26	9.41	6.54	5.41	4.79	4.38	4.10	3.89	3.73	3.60
27	9.34	6.49	5.36	4.74	4.34	4.06	3.85	3.69	3.56
28	9.28	6.44	5.32	4.70	4.30	4.02	3.81	3.65	3.52
29	9.23	6.40	5.28	4.66	4.26	3.98	3.77	3.61	3.48
30	9.18	6.35	5.24	4.62	4.23	3.95	3.74	3.58	3.45
40	8.83	6.07	4.98	4.37	3.99	3.71	3.51	3.35	3.22
60	8.49	5.79	4.73	4.14	3.76	3.49	3.29	3.13	3.01
120	8.18	5.54	4.50	3.92	3.55	3.28	3.09	2.93	2.81
∞	7.88	5.30	4.28	3.72	3.35	3.09	2.90	2.74	2.62

Source: From Beyer, W. H., *Handbook of Tables for Probability and Statistics, 2nd Edition,* CRC Press, Boca Raton, Florida, 1986. With permission.

10	12	15	20	24	30	40	60	120	∞
24224	24426	24630	24836	24940	25044	25148	25253	25359	25465
199.4	199.4	199.4	199.4	199.5	199.5	199.5	199.5	199.5	199.5
43.69	43.39	43.08	42.78	42.62	42.47	42.31	42.15	41.99	41.83
20.97	20.70	20.44	20.17	20.03	19.89	19.75	19.61	19.47	19.32
13.62	13.38	13.15	12.90	12.78	12.66	12.53	12.40	12.27	12.14
10.25	10.03	9.81	9.59	9.47	9.36	9.24	9.12	9.00	8.88
8.38	8.18	7.97	7.75	7.65	7.53	7.42	7.31	7.19	7.08
7.21	7.01	6.81	6.61	6.50	6.40	6.29	6.18	6.06	5.95
6.42	6.23	6.03	5.83	5.73	5.62	5.52	5.41	5.30	5.19
5.85	5.66	5.47	5.27	5.17	5.07	4.97	4.86	4.75	4.64
5.42	5.24	5.05	4.86	4.76	4.65	4.55	4.44	4.34	4.23
5.09	4.91	4.72	4.53	4.43	4.33	4.23	4.12	4.01	3.90
4.82	4.64	4.46	4.27	4.17	4.07	3.97	3.87	3.76	3.65
4.60	4.43	4.25	4.06	3.96	3.86	3.76	3.66	3.55	3.44
4.42	4.25	4.07	3.88	3.79	3.69	3.58	3.48	3.37	3.26
4.27	4.10	3.92	3.73	3.64	3.54	3.44	3.33	3.22	3.11
4.14	3.97	3.79	3.61	3.51	3.41	3.31	3.21	3.10	2.98
4.03	3.86	3.68	3.50	3.40	3.30	3.20	3.10	2.99	2.87
3.93	3.76	3.59	3.40	3.31	3.21	3.11	3.00	2.89	2.78
3.85	3.68	3.50	3.32	3.22	3.12	3.02	2.92	2.81	2.69
3.77	3.60	3.43	3.24	3.15	3.05	2.95	2.84	2.73	2.61
3.70	3.54	3.36	3.18	3.08	2.98	2.88	2.77	2.66	2.55
3.64	3.47	3.30	3.12	3.02	2.92	2.82	2.71	2.60	2.48
3.59	3.42	3.25	3.06	2.97	2.87	2.77	2.66	2.55	2.43
3.54	3.37	3.20	3.01	2.92	2.82	2.72	2.61	2.50	2.38
3.49	3.33	3.15	2.97	2.87	2.77	2.67	2.56	2.45	2.33
3.45	3.28	3.11	2.93	2.83	2.73	2.63	2.52	2.41	2.25
3.41	3.25	3.07	2.89	2.79	2.69	2.59	2.48	2.37	2.29
3.38	3.21	3.04	2.86	2.76	2.66	2.56	2.45	2.33	2.24
3.34	3.18	3.01	2.82	2.73	2.63	2.52	2.42	2.30	2.18
3.12	2.95	2.78	2.60	2.50	2.40	2.30	2.18	2.06	1.93
2.90	2.74	2.57	2.39	2.29	2.19	2.08	1.96	1.83	1.69
2.71	2.54	2.37	2.19	2.09	1.98	1.87	1.75	1.61	1.43
2.52	2.36	2.19	2.00	1.90	1.79	1.67	1.53	1.36	1.00

Table H

The F Distribution   *Continued*

d.f.D.: degrees of freedom, denominator	$\alpha = 0.01$								
	d.f.N.: Degrees of freedom, numerator								
	1	*2*	*3*	*4*	*5*	*6*	*7*	*8*	*9*
1	4052	4999.5	5403	5625	5764	5859	5928	5982	6022
2	98.50	99.00	99.17	99.25	99.30	99.33	99.36	99.37	99.39
3	34.12	30.82	29.46	28.71	28.24	27.91	27.67	27.49	27.35
4	21.20	18.00	16.69	15.98	15.52	15.21	14.98	14.80	14.66
5	16.26	13.27	12.06	11.39	10.97	10.67	10.46	10.29	10.16
6	13.75	10.92	9.78	9.15	8.75	8.47	8.26	8.10	7.98
7	12.25	9.55	8.45	7.85	7.46	7.19	6.99	6.84	6.72
8	11.26	8.65	7.59	7.01	6.63	6.37	6.18	6.03	5.91
9	10.56	8.02	6.99	6.42	6.06	5.80	5.61	5.47	5.35
10	10.04	7.56	6.55	5.99	5.64	5.39	5.20	5.06	4.94
11	9.65	7.21	6.22	5.67	5.32	5.07	4.89	4.74	4.63
12	9.33	6.93	5.95	5.41	5.06	4.82	4.64	4.50	4.39
13	9.07	6.70	5.74	5.21	4.86	4.62	4.44	4.30	4.19
14	8.86	6.51	5.56	5.04	4.69	4.46	4.28	4.14	4.03
15	8.68	6.36	5.42	4.89	4.56	4.32	4.14	4.00	3.89
16	8.53	6.23	5.29	4.77	4.44	4.20	4.03	3.89	3.78
17	8.40	6.11	5.18	4.67	4.34	4.10	3.93	3.79	3.68
18	8.29	6.01	5.09	4.58	4.25	4.01	3.84	3.71	3.60
19	8.18	5.93	5.01	4.50	4.17	3.94	3.77	3.63	3.52
20	8.10	5.85	4.94	4.43	4.10	3.87	3.70	3.56	3.46
21	8.02	5.78	4.87	4.37	4.04	3.81	3.64	3.51	3.40
22	7.95	5.72	4.82	4.31	3.99	3.76	3.59	3.45	3.35
23	7.88	5.66	4.76	4.26	3.94	3.71	3.54	3.41	3.30
24	7.82	5.61	4.72	4.22	3.90	3.67	3.50	3.36	3.26
25	7.77	5.57	4.68	4.18	3.85	3.63	3.46	3.32	3.22
26	7.72	5.53	4.64	4.14	3.82	3.59	3.42	3.29	3.18
27	7.68	5.49	4.60	4.11	3.78	3.56	3.39	3.26	3.15
28	7.64	5.45	4.57	4.07	3.75	3.53	3.36	3.23	3.12
29	7.60	5.42	4.54	4.04	3.73	3.50	3.33	3.20	3.09
30	7.56	5.39	4.51	4.02	3.70	3.47	3.30	3.17	3.07
40	7.31	5.18	4.31	3.83	3.51	3.29	3.12	2.99	2.89
60	7.08	4.98	4.13	3.65	3.34	3.12	2.95	2.82	2.72
120	6.85	4.79	3.95	3.48	3.17	2.96	2.79	2.66	2.56
∞	6.63	4.61	3.78	3.32	3.02	2.80	2.64	2.51	2.41

10	12	15	20	24	30	40	60	120	∞
6056	6106	6157	6209	6235	6261	6287	6313	6339	6366
99.40	99.42	99.43	99.45	99.46	99.47	99.47	99.48	99.49	99.50
27.23	27.05	26.87	26.69	26.60	26.50	26.41	26.32	26.22	26.13
14.55	14.37	14.20	14.02	13.93	13.84	13.75	13.65	13.56	13.46
10.05	9.89	9.72	9.55	9.47	9.38	9.29	9.20	9.11	9.02
7.87	7.72	7.56	7.40	7.31	7.23	7.14	7.06	6.97	6.88
6.62	6.47	6.31	6.16	6.07	5.99	5.91	5.82	5.74	5.65
5.81	5.67	5.52	5.36	5.28	5.20	5.12	5.03	4.95	4.86
5.26	5.11	4.96	4.81	4.73	4.65	4.57	4.48	4.40	4.31
4.85	4.71	4.56	4.41	4.33	4.25	4.17	4.08	4.00	3.91
4.54	4.40	4.25	4.10	4.02	3.94	3.86	3.78	3.69	3.60
4.30	4.16	4.01	3.86	3.78	3.70	3.62	3.54	3.45	3.36
4.10	3.96	3.82	3.66	3.59	3.51	3.43	3.34	3.25	3.17
3.94	3.80	3.66	3.51	3.43	3.35	3.27	3.18	3.09	3.00
3.80	3.67	3.52	3.37	3.29	3.21	3.13	3.05	2.96	2.87
3.69	3.55	3.41	3.26	3.18	3.10	3.02	2.93	2.84	2.75
3.59	3.46	3.31	3.16	3.08	3.00	2.92	2.83	2.75	2.65
3.51	3.37	3.23	3.08	3.00	2.92	2.84	2.75	2.66	2.57
3.43	3.30	3.15	3.00	2.92	2.84	2.76	2.67	2.58	2.49
3.37	3.23	3.09	2.94	2.86	2.78	2.69	2.61	2.52	2.42
3.31	3.17	3.03	2.88	2.80	2.72	2.64	2.55	2.46	2.36
3.26	3.12	2.98	2.83	2.75	2.67	2.58	2.50	2.40	2.31
3.21	3.07	2.93	2.78	2.70	2.62	2.54	2.45	2.35	2.26
3.17	3.03	2.89	2.74	2.66	2.58	2.49	2.40	2.31	2.21
3.13	2.99	2.85	2.70	2.62	2.54	2.45	2.36	2.27	2.17
3.09	2.96	2.81	2.66	2.58	2.50	2.42	2.33	2.23	2.13
3.06	2.93	2.78	2.63	2.55	2.47	2.38	2.29	2.20	2.10
3.03	2.90	2.75	2.60	2.52	2.44	2.35	2.26	2.17	2.06
3.00	2.87	2.73	2.57	2.49	2.41	2.33	2.23	2.14	2.03
2.98	2.84	2.70	2.55	2.47	2.39	2.30	2.21	2.11	2.01
2.80	2.66	2.52	2.37	2.29	2.20	2.11	2.02	1.92	1.80
2.63	2.50	2.35	2.20	2.12	2.03	1.94	1.84	1.73	1.60
2.47	2.34	2.19	2.03	1.95	1.86	1.76	1.66	1.53	1.38
2.32	2.18	2.04	1.88	1.79	1.70	1.59	1.47	1.32	1.00

Table H

The F Distribution *Continued*

d.f.D.: degrees of freedom, denominator	\(\alpha = 0.025 \) d.f.N.: Degrees of freedom, numerator								
	1	*2*	*3*	*4*	*5*	*6*	*7*	*8*	*9*
1	647.8	799.5	864.2	899.6	921.8	937.1	948.2	956.7	963.3
2	38.51	39.00	39.17	39.25	39.30	39.33	39.36	39.37	39.39
3	17.44	16.04	15.44	15.10	14.88	14.73	14.62	14.54	14.47
4	12.22	10.65	9.98	9.60	9.36	9.20	9.07	8.98	8.90
5	10.01	8.43	7.76	7.39	7.15	6.98	6.85	6.76	6.68
6	8.81	7.26	6.60	6.23	5.99	5.82	5.70	5.60	5.52
7	8.07	6.54	5.89	5.52	5.29	5.12	4.99	4.90	4.82
8	7.57	6.06	5.42	5.05	4.82	4.65	4.53	4.43	4.36
9	7.21	5.71	5.08	4.72	4.48	4.32	4.20	4.10	4.03
10	6.94	5.46	4.83	4.47	4.24	4.07	3.95	3.85	3.78
11	6.72	5.26	4.63	4.28	4.04	3.88	3.76	3.66	3.59
12	6.55	5.10	4.47	4.12	3.89	3.73	3.61	3.51	3.44
13	6.41	4.97	4.35	4.00	3.77	3.60	3.48	3.39	3.31
14	6.30	4.86	4.24	3.89	3.66	3.50	3.38	3.29	3.21
15	6.20	4.77	4.15	3.80	3.58	3.41	3.29	3.20	3.12
16	6.12	4.69	4.08	3.73	3.50	3.34	3.22	3.12	3.05
17	6.04	4.62	4.01	3.66	3.44	3.28	3.16	3.06	2.98
18	5.98	4.56	3.95	3.61	3.38	3.22	3.10	3.01	2.93
19	5.92	4.51	3.90	3.56	3.33	3.17	3.05	2.96	2.88
20	5.87	4.46	3.86	3.51	3.29	3.13	3.01	2.91	2.84
21	5.83	4.42	3.82	3.48	3.25	3.09	2.97	2.87	2.80
22	5.79	4.38	3.78	3.44	3.22	3.05	2.93	2.84	2.76
23	5.75	4.35	3.75	3.41	3.18	3.02	2.90	2.81	2.73
24	5.72	4.32	3.72	3.38	3.15	2.99	2.87	2.78	2.70
25	5.69	4.29	3.69	3.35	3.13	2.97	2.85	2.75	2.68
26	5.66	4.27	3.67	3.33	3.10	2.94	2.82	2.73	2.65
27	5.63	4.24	3.65	3.31	3.08	2.92	2.80	2.71	2.63
28	5.61	4.22	3.63	3.29	3.06	2.90	2.78	2.69	2.61
29	5.59	4.20	3.61	3.27	3.04	2.88	2.76	2.67	2.59
30	5.57	4.18	3.59	3.25	3.03	2.87	2.75	2.65	2.57
40	5.42	4.05	3.46	3.13	2.90	2.74	2.62	2.53	2.45
60	5.29	3.93	3.34	3.01	2.79	2.63	2.51	2.41	2.33
120	5.15	3.80	3.23	2.89	2.67	2.52	2.39	2.30	2.22
∞	5.02	3.69	3.12	2.79	2.57	2.41	2.29	2.19	2.11

10	12	15	20	24	30	40	60	120	∞
968.6	976.7	984.9	993.1	997.2	1001	1006	1010	1014	1018
39.40	39.41	39.43	39.45	39.46	39.46	39.47	39.48	39.49	39.50
14.42	14.34	14.25	14.17	14.12	14.08	14.04	13.99	13.95	13.90
8.84	8.75	8.66	8.56	8.51	8.46	8.41	8.36	8.31	8.26
6.62	6.52	6.43	6.33	6.28	6.23	6.18	6.12	6.07	6.02
5.46	5.37	5.27	5.17	5.12	5.07	5.01	4.96	4.90	4.85
4.76	4.67	4.57	4.47	4.42	4.36	4.31	4.25	4.20	4.14
4.30	4.20	4.10	4.00	3.95	3.89	3.84	3.78	3.73	3.67
3.96	3.87	3.77	3.67	3.61	3.56	3.51	3.45	3.39	3.33
3.72	3.62	3.52	3.42	3.37	3.31	3.26	3.20	3.14	3.08
3.53	3.43	3.33	3.23	3.17	3.12	3.06	3.00	2.94	2.88
3.37	3.28	3.18	3.07	3.02	2.96	2.91	2.85	2.79	2.72
3.25	3.15	3.05	2.95	2.89	2.84	2.78	2.72	2.66	2.60
3.15	3.05	2.95	2.84	2.79	2.73	2.67	2.61	2.55	2.49
3.06	2.96	2.86	2.76	2.70	2.64	2.59	2.52	2.46	2.40
2.99	2.89	2.79	2.68	2.63	2.57	2.51	2.45	2.38	2.32
2.92	2.82	2.72	2.62	2.56	2.50	2.44	2.38	2.32	2.25
2.87	2.77	2.67	2.56	2.50	2.44	2.38	2.32	2.26	2.19
2.82	2.72	2.62	2.51	2.45	2.39	2.33	2.27	2.20	2.13
2.77	2.68	2.57	2.46	2.41	2.35	2.29	2.22	2.16	2.09
2.73	2.64	2.53	2.42	2.37	2.31	2.25	2.18	2.11	2.04
2.70	2.60	2.50	2.39	2.33	2.27	2.21	2.14	2.08	2.00
2.67	2.57	2.47	2.36	2.30	2.24	2.18	2.11	2.04	1.97
2.64	2.54	2.44	2.33	2.27	2.21	2.15	2.08	2.01	1.94
2.61	2.51	2.41	2.30	2.24	2.18	2.12	2.05	1.98	1.91
2.59	2.49	2.39	2.28	2.22	2.16	2.09	2.03	1.95	1.88
2.57	2.47	2.36	2.25	2.19	2.13	2.07	2.00	1.93	1.85
2.55	2.45	2.34	2.23	2.17	2.11	2.05	1.98	1.91	1.83
2.53	2.43	2.32	2.21	2.15	2.09	2.03	1.96	1.89	1.81
2.51	2.41	2.31	2.20	2.14	2.07	2.01	1.94	1.87	1.79
2.39	2.29	2.18	2.07	2.01	1.94	1.88	1.80	1.72	1.64
2.27	2.17	2.06	1.94	1.88	1.82	1.74	1.67	1.58	1.48
2.16	2.05	1.94	1.82	1.76	1.69	1.61	1.53	1.43	1.31
2.05	1.94	1.83	1.71	1.64	1.57	1.48	1.39	1.27	1.00

Table H

The F Distribution *Continued*

d.f.D.: degrees of freedom, denominator	$\alpha = 0.05$								
	d.f.N.: Degrees of freedom, numerator								
	1	2	3	4	5	6	7	8	9
1	161.4	199.5	215.7	224.6	230.2	234.0	236.8	238.9	240.5
2	18.51	19.00	19.16	19.25	19.30	19.33	19.35	19.37	19.38
3	10.13	9.55	9.28	9.12	9.01	8.94	8.89	8.85	8.81
4	7.71	6.94	6.59	6.39	6.26	6.16	6.09	6.04	6.00
5	6.61	5.79	5.41	5.19	5.05	4.95	4.88	4.82	4.77
6	5.99	5.14	4.76	4.53	4.39	4.28	4.21	4.15	4.10
7	5.59	4.74	4.35	4.12	3.97	3.87	3.79	3.73	3.68
8	5.32	4.46	4.07	3.84	3.69	3.58	3.50	3.44	3.39
9	5.12	4.26	3.86	3.63	3.48	3.37	3.29	3.23	3.18
10	4.96	4.10	3.71	3.48	3.33	3.22	3.14	3.07	3.02
11	4.84	3.98	3.59	3.36	3.20	3.09	3.01	2.95	2.90
12	4.75	3.89	3.49	3.26	3.11	3.00	2.91	2.85	2.80
13	4.67	3.81	3.41	3.18	3.03	2.92	2.83	2.77	2.71
14	4.60	3.74	3.34	3.11	2.96	2.85	2.76	2.70	2.65
15	4.54	3.68	3.29	3.06	2.90	2.79	2.71	2.64	2.59
16	4.49	3.63	3.24	3.01	2.85	2.74	2.66	2.59	2.54
17	4.45	3.59	3.20	2.96	2.81	2.70	2.61	2.55	2.49
18	4.41	3.55	3.16	2.93	2.77	2.66	2.58	2.51	2.46
19	4.38	3.52	3.13	2.90	2.74	2.63	2.54	2.48	2.42
20	4.35	3.49	3.10	2.87	2.71	2.60	2.51	2.45	2.39
21	4.32	3.47	3.07	2.84	2.68	2.57	2.49	2.42	2.37
22	4.30	3.44	3.05	2.82	2.66	2.55	2.46	2.40	2.34
23	4.28	3.42	3.03	2.80	2.64	2.53	2.44	2.37	2.32
24	4.26	3.40	3.01	2.78	2.62	2.51	2.42	2.36	2.30
25	4.24	3.39	2.99	2.76	2.60	2.49	2.40	2.34	2.28
26	4.23	3.37	2.98	2.74	2.59	2.47	2.39	2.32	2.27
27	4.21	3.35	2.96	2.73	2.57	2.46	2.37	2.31	2.25
28	4.20	3.34	2.95	2.71	2.56	2.45	2.36	2.29	2.24
29	4.18	3.33	2.93	2.70	2.55	2.43	2.35	2.28	2.22
30	4.17	3.32	2.92	2.69	2.53	2.42	2.33	2.27	2.21
40	4.08	3.23	2.84	2.61	2.45	2.34	2.25	2.18	2.12
60	4.00	3.15	2.76	2.53	2.37	2.25	2.17	2.10	2.04
120	3.92	3.07	2.68	2.45	2.29	2.17	2.09	2.02	1.96
∞	3.84	3.00	2.60	2.37	2.21	2.10	2.01	1.94	1.88

10	12	15	20	24	30	40	60	120	∞
241.9	243.9	245.9	248.0	249.1	250.1	251.1	252.2	253.3	254.3
19.40	19.41	19.43	19.45	19.45	19.46	19.47	19.48	19.49	19.50
8.79	8.74	8.70	8.66	8.64	8.62	8.59	8.57	8.55	8.53
5.96	5.91	5.86	5.80	5.77	5.75	5.72	5.69	5.66	5.63
4.74	4.68	4.62	4.56	4.53	4.50	4.46	4.43	4.40	4.36
4.06	4.00	3.94	3.87	3.84	3.81	3.77	3.74	3.70	3.67
3.64	3.57	3.51	3.44	3.41	3.38	3.34	3.30	3.27	3.23
3.35	3.28	3.22	3.15	3.12	3.08	3.04	3.01	2.97	2.93
3.14	3.07	3.01	2.94	2.90	2.86	2.83	2.79	2.75	2.71
2.98	2.91	2.85	2.77	2.74	2.70	2.66	2.62	2.58	2.54
2.85	2.79	2.72	2.65	2.61	2.57	2.53	2.49	2.45	2.40
2.75	2.69	2.62	2.54	2.51	2.47	2.43	2.38	2.34	2.30
2.67	2.60	2.53	2.46	2.42	2.38	2.34	2.30	2.25	2.21
2.60	2.53	2.46	2.39	2.35	2.31	2.27	2.22	2.18	2.13
2.54	2.48	2.40	2.33	2.29	2.25	2.20	2.16	2.11	2.07
2.49	2.42	2.35	2.28	2.24	2.19	2.15	2.11	2.06	2.01
2.45	2.38	2.31	2.23	2.19	2.15	2.10	2.06	2.01	1.96
2.41	2.34	2.27	2.19	2.15	2.11	2.06	2.02	1.97	1.92
2.38	2.31	2.23	2.16	2.11	2.07	2.03	1.98	1.93	1.88
2.35	2.28	2.20	2.12	2.08	2.04	1.99	1.95	1.90	1.84
2.32	2.25	2.18	2.10	2.05	2.01	1.96	1.92	1.87	1.81
2.30	2.23	2.15	2.07	2.03	1.98	1.94	1.89	1.84	1.78
2.27	2.20	2.13	2.05	2.01	1.96	1.91	1.86	1.81	1.76
2.25	2.18	2.11	2.03	1.98	1.94	1.89	1.84	1.79	1.73
2.24	2.16	2.09	2.01	1.96	1.92	1.87	1.82	1.77	1.71
2.22	2.15	2.07	1.99	1.95	1.90	1.85	1.80	1.75	1.69
2.20	2.13	2.06	1.97	1.93	1.88	1.84	1.79	1.73	1.67
2.19	2.12	2.04	1.96	1.91	1.87	1.82	1.77	1.71	1.65
2.18	2.10	2.03	1.94	1.90	1.85	1.81	1.75	1.70	1.64
2.16	2.09	2.01	1.93	1.89	1.84	1.79	1.74	1.68	1.62
2.08	2.00	1.92	1.84	1.79	1.74	1.69	1.64	1.58	1.51
1.99	1.92	1.84	1.75	1.70	1.65	1.59	1.53	1.47	1.39
1.91	1.83	1.75	1.66	1.61	1.55	1.50	1.43	1.35	1.25
1.83	1.75	1.67	1.57	1.52	1.46	1.39	1.32	1.22	1.00

Table H

The *F* Distribution   *Continued*

d.f.D.: degrees of freedom, denominator	α = 0.10 d.f.N.: Degrees of freedom, numerator								
	1	*2*	*3*	*4*	*5*	*6*	*7*	*8*	*9*
1	39.86	49.50	53.59	55.83	57.24	58.20	58.91	59.44	59.86
2	8.53	9.00	9.16	9.24	9.29	9.33	9.35	9.37	9.38
3	5.54	5.46	5.39	5.34	5.31	5.28	5.27	5.25	5.24
4	4.54	4.32	4.19	4.11	4.05	4.01	3.98	3.95	3.94
5	4.06	3.78	3.62	3.52	3.45	3.40	3.37	3.34	3.32
6	3.78	3.46	3.29	3.18	3.11	3.05	3.01	2.98	2.96
7	3.59	3.26	3.07	2.96	2.88	2.83	2.78	2.75	2.72
8	3.46	3.11	2.92	2.81	2.73	2.67	2.62	2.59	2.56
9	3.36	3.01	2.81	2.69	2.61	2.55	2.51	2.47	2.44
10	3.29	2.92	2.73	2.61	2.52	2.46	2.41	2.38	2.35
11	3.23	2.86	2.66	2.54	2.45	2.39	2.34	2.30	2.27
12	3.18	2.81	2.61	2.48	2.39	2.33	2.28	2.24	2.21
13	3.14	2.76	2.56	2.43	2.35	2.28	2.23	2.20	2.16
14	3.10	2.73	2.52	2.39	2.31	2.24	2.19	2.15	2.12
15	3.07	2.70	2.49	2.36	2.27	2.21	2.16	2.12	2.09
16	3.05	2.67	2.46	2.33	2.24	2.18	2.13	2.09	2.06
17	3.03	2.64	2.44	2.31	2.22	2.15	2.10	2.06	2.03
18	3.01	2.62	2.42	2.29	2.20	2.13	2.08	2.04	2.00
19	2.99	2.61	2.40	2.27	2.18	2.11	2.06	2.02	1.98
20	2.97	2.59	2.38	2.25	2.16	2.09	2.04	2.00	1.96
21	2.96	2.57	2.36	2.23	2.14	2.08	2.02	1.98	1.95
22	2.95	2.56	2.35	2.22	2.13	2.06	2.01	1.97	1.93
23	2.94	2.55	2.34	2.21	2.11	2.05	1.99	1.95	1.92
24	2.93	2.54	2.33	2.19	2.10	2.04	1.98	1.94	1.91
25	2.92	2.53	2.32	2.18	2.09	2.02	1.97	1.93	1.89
26	2.91	2.52	2.31	2.17	2.08	2.01	1.96	1.92	1.88
27	2.90	2.51	2.30	2.17	2.07	2.00	1.95	1.91	1.87
28	2.89	2.50	2.29	2.16	2.06	2.00	1.94	1.90	1.87
29	2.89	2.50	2.28	2.15	2.06	1.99	1.93	1.89	1.86
30	2.88	2.49	2.28	2.14	2.05	1.98	1.93	1.88	1.85
40	2.84	2.44	2.23	2.09	2.00	1.93	1.87	1.83	1.79
60	2.79	2.39	2.18	2.04	1.95	1.87	1.82	1.77	1.74
120	2.75	2.35	2.13	1.99	1.90	1.82	1.77	1.72	1.68
∞	2.71	2.30	2.08	1.94	1.85	1.77	1.72	1.67	1.63

10	12	15	20	24	30	40	60	120	∞
60.19	60.71	61.22	61.74	62.00	62.26	62.53	62.79	63.06	63.33
9.39	9.41	9.42	9.44	9.45	9.46	9.47	9.47	9.48	9.49
5.23	5.22	5.20	5.18	5.18	5.17	5.16	5.15	5.14	5.13
3.92	3.90	3.87	3.84	3.83	3.82	3.80	3.79	3.78	3.76
3.30	3.27	3.24	3.21	3.19	3.17	3.16	3.14	3.12	3:10
2.94	2.90	2.87	2.84	2.82	2.80	2.78	2.76	2.74	2.72
2.70	2.67	2.63	2.59	2.58	2.56	2.54	2.51	2.49	2.47
2.54	2.50	2.46	2.42	2.40	2.38	2.36	2.34	2.32	2.29
2.42	2.38	2.34	2.30	2.28	2.25	2.23	2.21	2.18	2.16
2.32	2.28	2.24	2.20	2.18	2.16	2.13	2.11	2.08	2.06
2.25	2.21	2.17	2.12	2.10	2.08	2.05	2.03	2.00	1.97
2.19	2.15	2.10	2.06	2.04	2.01	1.99	1.96	1.93	1.90
2.14	2.10	2.05	2.01	1.98	1.96	1.93	1.90	1.88	1.85
2.10	2.05	2.01	1.96	1.94	1.91	1.89	1.86	1.83	1.80
2.06	2.02	1.97	1.92	1.90	1.87	1.85	1.82	1.79	1.76
2.03	1.99	1.94	1.89	1.87	1.84	1.81	1.78	1.75	1.72
2.00	1.96	1.91	1.86	1.84	1.81	1.78	1.75	1.72	1.69
1.98	1.93	1.89	1.84	1.81	1.78	1.75	1.72	1.69	1.66
1.96	1.91	1.86	1.81	1.79	1.76	1.73	1.70	1.67	1.63
1.94	1.89	1.84	1.79	1.77	1.74	1.71	1.68	1.64	1.61
1.92	1.87	1.83	1.78	1.75	1.72	1.69	1.66	1.62	1.59
1.90	1.86	1.81	1.76	1.73	1.70	1.67	1.64	1.60	1.57
1.89	1.84	1.80	1.74	1.72	1.69	1.66	1.62	1.59	1.55
1.88	1.83	1.78	1.73	1.70	1.67	1.64	1.61	1.57	1.53
1.87	1.82	1.77	1.72	1.69	1.66	1.63	1.59	1.56	1.52
1.86	1.81	1.76	1.71	1.68	1.65	1.61	1.58	1.54	1.50
1.85	1.80	1.75	1.70	1.67	1.64	1.60	1.57	1.53	1.49
1.84	1.79	1.74	1.69	1.66	1.63	1.59	1.56	1.52	1.48
1.83	1.78	1.73	1.68	1.65	1.62	1.58	1.55	1.51	1.47
1.82	1.77	1.72	1.67	1.64	1.61	1.57	1.54	1.50	1.46
1.76	1.71	1.66	1.61	1.57	1.54	1.51	1.47	1.42	1.38
1.71	1.66	1.60	1.54	1.51	1.48	1.44	1.40	1.35	1.29
1.65	1.60	1.55	1.48	1.45	1.41	1.37	1.32	1.26	1.19
1.60	1.55	1.49	1.42	1.38	1.34	1.30	1.24	1.17	1.00

Table I

Critical Values for the PPMC

Reject H_0: $\rho = 0$ if the absolute value of r is greater than the value given in the table. The values are for a two-tailed test; d.f. $= n - 2$.

d.f.	$\alpha = 0.05$	$\alpha = 0.01$
1	0.999	0.999
2	0.950	0.999
3	0.878	0.959
4	0.811	0.917
5	0.754	0.875
6	0.707	0.834
7	0.666	0.798
8	0.632	0.765
9	0.602	0.735
10	0.576	0.708
11	0.553	0.684
12	0.532	0.661
13	0.514	0.641
14	0.497	0.623
15	0.482	0.606
16	0.468	0.590
17	0.456	0.575
18	0.444	0.561
19	0.433	0.549
20	0.423	0.537
25	0.381	0.487
30	0.349	0.449
35	0.325	0.418
40	0.304	0.393
45	0.288	0.372
50	0.273	0.354
60	0.250	0.325
70	0.232	0.302
80	0.217	0.283
90	0.205	0.267
100	0.195	0.254

Source: From Beyer, W. H., *Handbook of Tables for Probability and Statistics, 2nd Edition,* CRC Press, Boca Raton, Florida, 1986. With permission.

Table J

Critical Values for the Sign Test

Reject the null hypothesis if the smaller number of + or − signs is less than or equal to the value in the table.

n	One-tailed, $\alpha =$ 0.005 Two-tailed, $\alpha =$ 0.01	0.01 0.02	0.025 0.05	0.05 0.10	
8	0	0	0	1	
9	0	0	1	1	
10	0	0	1	1	
11	0	1	1	2	
12	1	1	2	2	
13	1	1	2	3	
14	1	2	3	3	
15	2	2	3	3	
16	2	2	3	4	*Note:* Table J is for
17	2	3	4	4	one-tailed or two-
18	3	3	4	5	tailed tests. The
19	3	4	4	5	term n represents the
20	3	4	5	5	total number of +
21	4	4	5	6	and − signs. The
22	4	5	5	6	test value is the
23	4	5	6	7	number of less
24	5	5	6	7	frequent signs.
25	5	6	6	7	

Source: From Beyer, W. H., *Handbook of Tables for Probability and Statistics, 2nd Edition,* CRC Press, Boca Raton, Florida, 1986. With permission.

Table K

Critical Values for the Wilcoxon Signed-Rank Test

Reject the null hypothesis if the test value is less than or equal to the value given in the table.

n	One-tailed, $\alpha = 0.05$ Two-tailed, $\alpha = 0.10$	0.025 0.05	0.01 0.02	0.005 0.01
5	1			
6	2	1		
7	4	2	0	
8	6	4	2	0
9	8	6	3	2
10	11	8	5	3
11	14	11	7	5
12	17	14	10	7
13	21	17	13	10
14	26	21	16	13
15	30	25	20	16
16	36	30	24	19
17	41	35	28	23
18	47	40	33	28
19	54	46	38	32
20	60	52	43	37
21	68	59	49	43
22	75	66	56	49
23	83	73	62	55
24	92	81	69	61
25	101	90	77	68
26	110	98	85	76
27	120	107	93	84
28	130	117	102	92
29	141	127	111	100
30	152	137	120	109

Source: From Beyer, W. H., *Handbook of Tables for Probability and Statistics. 2nd Edition,* CRC Press, Boca Raton, Florida, 1986. With permission.

Table L

Critical Values for the Rank Correlation Coefficient

Reject H_0: $\rho = 0$ if the absolute value of r_S is greater than the value given in the table.

n	$\alpha = 0.10$	$\alpha = 0.05$	$\alpha = 0.02$	$\alpha = 0.01$
5	0.900	—	—	—
6	0.829	0.886	0.943	—
7	0.714	0.786	0.893	0.929
8	0.643	0.738	0.833	0.881
9	0.600	0.700	0.783	0.833
10	0.564	0.648	0.745	0.794
11	0.536	0.618	0.709	0.818
12	0.497	0.591	0.703	0.780
13	0.475	0.566	0.673	0.745
14	0.457	0.545	0.646	0.716
15	0.441	0.525	0.623	0.689
16	0.425	0.507	0.601	0.666
17	0.412	0.490	0.582	0.645
18	0.399	0.476	0.564	0.625
19	0.388	0.462	0.549	0.608
20	0.377	0.450	0.534	0.591
21	0.368	0.438	0.521	0.576
22	0.359	0.428	0.508	0.562
23	0.351	0.418	0.496	0.549
24	0.343	0.409	0.485	0.537
25	0.336	0.400	0.475	0.526
26	0.329	0.392	0.465	0.515
27	0.323	0.385	0.456	0.505
28	0.317	0.377	0.488	0.496
29	0.311	0.370	0.440	0.487
30	0.305	0.364	0.432	0.478

Source: From Beyer, W. H., *Handbook of Tables for Probability and Statistics, 2nd Edition,* CRC Press, Boca Raton, Florida, 1986. With permission.

Table M

Critical Values for the Number of Runs

This table gives the critical values at $\alpha = 0.05$ for a two-tailed test. Reject the null hypothesis if the number of runs is less than or equal to the smaller value or greater than or equal to the larger value.

Value of n_1	Value of n_2																		
	2	3	4	5	6	7	8	9	10	11	12	13	14	15	16	17	18	19	20
2	1	1	1	1	1	1	1	1	1	1	2	2	2	2	2	2	2	2	2
	6	6	6	6	6	6	6	6	6	6	6	6	6	6	6	6	6	6	6
3	1	1	1	1	2	2	2	2	2	2	2	2	2	3	3	3	3	3	3
	6	8	8	8	8	8	8	8	8	8	8	8	8	8	8	8	8	8	8
4	1	1	1	2	2	2	3	3	3	3	3	3	3	3	4	4	4	4	4
	6	8	9	9	9	10	10	10	10	10	10	10	10	10	10	10	10	10	10
5	1	1	2	2	3	3	3	3	3	4	4	4	4	4	4	4	5	5	5
	6	8	9	10	10	11	11	12	12	12	12	12	12	12	12	12	12	12	12
6	1	2	2	3	3	3	3	4	4	4	4	5	5	5	5	5	5	6	6
	6	8	9	10	11	12	12	13	13	13	13	14	14	14	14	14	14	14	14
7	1	2	2	3	3	3	4	4	5	5	5	5	5	6	6	6	6	6	6
	6	8	10	11	12	13	13	14	14	14	14	15	15	15	16	16	16	16	16
8	1	2	3	3	3	4	4	5	5	5	6	6	6	6	6	7	7	7	7
	6	8	10	11	12	13	14	14	15	15	16	16	16	16	17	17	17	17	17
9	1	2	3	3	4	4	5	5	5	6	6	6	7	7	7	7	8	8	8
	6	8	10	12	13	14	14	15	16	16	16	17	17	18	18	18	18	18	18
10	1	2	3	3	4	5	5	5	6	6	7	7	7	7	8	8	8	8	9
	6	8	10	12	13	14	15	16	16	17	17	18	18	18	19	19	19	20	20
11	1	2	3	4	4	5	5	6	6	7	7	7	8	8	8	9	9	9	9
	6	8	10	12	13	14	15	16	17	17	18	19	19	19	20	20	20	21	21
12	2	2	3	4	4	5	6	6	7	7	7	8	8	8	9	9	9	10	10
	6	8	10	12	13	14	16	16	17	18	19	19	20	20	21	21	21	22	22

Table M

Critical Values for the Number of Runs *Continued*

This table gives the critical values at $\alpha = 0.05$ for a two-tailed test. Reject the null hypothesis if the number of runs is less than or equal to the smaller value or greater than or equal to the larger value.

Value of n_1	Value of n_2																		
	2	3	4	5	6	7	8	9	10	11	12	13	14	15	16	17	18	19	20
13	2	2	3	4	5	5	6	6	7	7	8	8	9	9	9	10	10	10	10
	6	8	10	12	14	15	16	17	18	19	19	20	20	21	21	22	22	23	23
14	2	2	3	4	5	5	6	7	7	8	8	9	9	9	10	10	10	11	11
	6	8	10	12	14	15	16	17	18	19	20	20	21	22	22	23	23	23	24
15	2	3	3	4	5	6	6	7	7	8	8	9	9	10	10	11	11	11	12
	6	8	10	12	14	15	16	18	18	19	20	21	22	22	23	23	24	24	25
16	2	3	4	4	5	6	6	7	8	8	9	9	10	10	11	11	11	12	12
	6	8	10	12	14	16	17	18	19	20	21	21	22	23	23	24	25	25	25
17	2	3	4	4	5	6	7	7	8	9	9	10	10	11	11	11	12	12	13
	6	8	10	12	14	16	17	18	19	20	21	22	23	23	24	25	25	26	26
18	2	3	4	5	5	6	7	8	8	9	9	10	10	11	11	12	12	13	13
	6	8	10	12	14	16	17	18	19	20	21	22	23	24	25	25	26	26	27
19	2	3	4	5	6	6	7	8	8	9	10	10	11	11	12	12	13	13	13
	6	8	10	12	14	16	17	18	20	21	22	23	23	24	25	26	26	27	27
20	2	3	4	5	6	6	7	8	9	9	10	10	11	12	12	13	13	13	14
	6	8	10	12	14	16	17	18	20	21	22	23	24	25	25	26	27	27	28

Source: Adapted from C. Eisenhardt and F. Swed, "Tables for Testing Randomness of Grouping in a Sequence of Alternatives," *The Annals of Statistics,* 14 (1943) 83–86. Reprinted with permission of the Institute of Mathematical Statistics and of the Benjamin/Cummings Publishing Company, in whose publication, *Elementary Statistics,* 3rd edition (1989), by Mario F. Triola, this table appears.

268

Table N

Critical Values for the Tukey Test

$\alpha = 0.01$

k / ν	2	3	4	5	6	7	8	9	10
1	90.03	135.0	164.3	185.6	202.2	215.8	227.2	237.0	245.6
2	14.04	19.02	22.29	24.72	26.63	28.20	29.53	30.68	31.69
3	8.26	10.62	12.17	13.33	14.24	15.00	15.64	16.20	16.69
4	6.51	8.12	9.17	9.96	10.58	11.10	11.55	11.93	12.27
5	5.70	6.98	7.80	8.42	8.91	9.32	9.67	9.97	10.24
6	5.24	6.33	7.03	7.56	7.97	8.32	8.61	8.87	9.10
7	4.95	5.92	6.54	7.01	7.37	7.68	7.94	8.17	8.37
8	4.75	5.64	6.20	6.62	6.96	7.24	7.47	7.68	7.86
9	4.60	5.43	5.96	6.35	6.66	6.91	7.13	7.33	7.49
10	4.48	5.27	5.77	6.14	6.43	6.67	6.87	7.05	7.21
11	4.39	5.15	5.62	5.97	6.25	6.48	6.67	6.84	6.99
12	4.32	5.05	5.50	5.84	6.10	6.32	6.51	6.67	6.81
13	4.26	4.96	5.40	5.73	5.98	6.19	6.37	6.53	6.67
14	4.21	4.89	5.32	5.63	5.88	6.08	6.26	6.41	6.54
15	4.17	4.84	5.25	5.56	5.80	5.99	6.16	6.31	6.44
16	4.13	4.79	5.19	5.49	5.72	5.92	6.08	6.22	6.35
17	4.10	4.74	5.14	5.43	5.66	5.85	6.01	6.15	6.27
18	4.07	4.70	5.09	5.38	5.60	5.79	5.94	6.08	6.20
19	4.05	4.67	5.05	5.33	5.55	5.73	5.89	6.02	6.14
20	4.02	4.64	5.02	5.29	5.51	5.69	5.84	5.97	6.09
24	3.96	4.55	4.91	5.17	5.37	5.54	5.69	5.81	5.92
30	3.89	4.45	4.80	5.05	5.24	5.40	5.54	5.65	5.76
40	3.82	4.37	4.70	4.93	5.11	5.26	5.39	5.50	5.60
60	3.76	4.28	4.59	4.82	4.99	5.13	5.25	5.36	5.45
120	3.70	4.20	4.50	4.71	4.87	5.01	5.12	5.21	5.30
∞	3.64	4.12	4.40	4.60	4.76	4.88	4.99	5.08	5.16

Source: From Beyer, W. H., *Handbook of Tables for Probability and Statistics, 2nd Edition,* CRC Press, Boca Raton, Florida, 1986. With permission.

Table N

Critical Values for the Tukey Test *Continued*

$\alpha = 0.01$

v \ k	11	12	13	14	15	16	17	18	19	20
1	253.2	260.0	266.2	271.8	277.0	281.8	286.3	290.4	294.3	298.0
2	32.59	33.40	34.13	34.81	35.43	36.00	36.53	37.03	37.50	37.95
3	17.13	17.53	17.89	18.22	18.52	18.81	19.07	19.32	19.55	19.77
4	12.57	12.84	13.09	13.32	13.53	13.73	13.91	14.08	14.24	14.40
5	10.48	10.70	10.89	11.08	11.24	11.40	11.55	11.68	11.81	11.93
6	9.30	9.48	9.65	9.81	9.95	10.08	10.21	10.32	10.43	10.54
7	8.55	8.71	8.86	9.00	9.12	9.24	9.35	9.46	9.55	9.65
8	8.03	8.18	8.31	8.44	8.55	8.66	8.76	8.85	8.94	9.03
9	7.65	7.78	7.91	8.03	8.13	8.23	8.33	8.41	8.49	8.57
10	7.36	7.49	7.60	7.71	7.81	7.91	7.99	8.08	8.15	8.23
11	7.13	7.25	7.36	7.46	7.56	7.65	7.73	7.81	7.88	7.95
12	6.94	7.06	7.17	7.26	7.36	7.44	7.52	7.59	7.66	7.73
13	6.79	6.90	7.01	7.10	7.19	7.27	7.35	7.42	7.48	7.55
14	6.66	6.77	6.87	6.96	7.05	7.13	7.20	7.27	7.33	7.39
15	6.55	6.66	6.76	6.84	6.93	7.00	7.07	7.14	7.20	7.26
16	6.46	6.56	6.66	6.74	6.82	6.90	6.97	7.03	7.09	7.15
17	6.38	6.48	6.57	6.66	6.73	6.81	6.87	6.94	7.00	7.05
18	6.31	6.41	6.50	6.58	6.65	6.73	6.79	6.85	6.91	6.97
19	6.25	6.34	6.43	6.51	6.58	6.65	6.72	6.78	6.84	6.89
20	6.19	6.28	6.37	6.45	6.52	6.59	6.65	6.71	6.77	6.82
24	6.02	6.11	6.19	6.26	6.33	6.39	6.45	6.51	6.56	6.61
30	5.85	5.93	6.01	6.08	6.14	6.20	6.26	6.31	6.36	6.41
40	5.69	5.76	5.83	5.90	5.96	6.02	6.07	6.12	6.16	6.21
60	5.53	5.60	5.67	5.73	5.78	5.84	5.89	5.93	5.97	6.01
120	5.37	5.44	5.50	5.56	5.61	5.66	5.71	5.75	5.79	5.83
∞	5.23	5.29	5.35	5.40	5.45	5.49	5.54	5.57	5.61	5.65

Table N

Critical Values for the Tukey Test *Continued*

$\alpha = 0.05$

v \ k	2	3	4	5	6	7	8	9	10
1	17.97	26.98	32.82	37.08	40.41	43.12	45.40	47.36	49.07
2	6.08	8.33	9.80	10.88	11.74	12.44	13.03	13.54	13.99
3	4.50	5.91	6.82	7.50	8.04	8.48	8.85	9.18	9.46
4	3.93	5.04	5.76	6.29	6.71	7.05	7.35	7.60	7.83
5	3.64	4.60	5.22	5.67	6.03	6.33	6.58	6.80	6.99
6	3.46	4.34	4.90	5.30	5.63	5.90	6.12	6.32	6.49
7	3.34	4.16	4.68	5.06	5.36	5.61	5.82	6.00	6.16
8	3.26	4.04	4.53	4.89	5.17	5.40	5.60	5.77	5.92
9	3.20	3.95	4.41	4.76	5.02	5.24	5.43	5.59	5.74
10	3.15	3.88	4.33	4.65	4.91	5.12	5.30	5.46	5.60
11	3.11	3.82	4.26	4.57	4.82	5.03	5.20	5.35	5.49
12	3.08	3.77	4.20	4.51	4.75	4.95	5.12	5.27	5.39
13	3.06	3.73	4.15	4.45	4.69	4.88	5.05	5.19	5.32
14	3.03	3.70	4.11	4.41	4.64	4.83	4.99	5.13	5.25
15	3.01	3.67	4.08	4.37	4.59	4.78	4.94	5.08	5.20
16	3.00	3.65	4.05	4.33	4.56	4.74	4.90	5.03	5.15
17	2.98	3.63	4.02	4.30	4.52	4.70	4.86	4.99	5.11
18	2.97	3.61	4.00	4.28	4.49	4.67	4.82	4.96	5.07
19	2.96	3.59	3.98	4.25	4.47	4.65	4.79	4.92	5.04
20	2.95	3.58	3.96	4.23	4.45	4.62	4.77	4.90	5.01
24	2.92	3.53	3.90	4.17	4.37	4.54	4.68	4.81	4.92
30	2.89	3.49	3.85	4.10	4.30	4.46	4.60	4.72	4.82
40	2.86	3.44	3.79	4.04	4.23	4.39	4.52	4.63	4.73
60	2.83	3.40	3.74	3.98	4.16	4.31	4.44	4.55	4.65
120	2.80	3.36	3.68	3.92	4.10	4.24	4.36	4.47	4.56
∞	2.77	3.31	3.63	3.86	4.03	4.17	4.29	4.39	4.47

Table N

Critical Values for the Tukey Test *Continued*

$\alpha = 0.05$

v \ k	11	12	13	14	15	16	17	18	19	20
1	50.59	51.96	53.20	54.33	55.36	56.32	57.22	58.04	58.83	59.56
2	14.39	14.75	15.08	15.38	15.65	15.91	16.14	16.37	16.57	16.77
3	9.72	9.95	10.15	10.35	10.53	10.69	10.84	10.98	11.11	11.24
4	8.03	8.21	8.37	8.52	8.66	8.79	8.91	9.03	9.13	9.23
5	7.17	7.32	7.47	7.60	7.72	7.83	7.93	8.03	8.12	8.21
6	6.65	6.79	6.92	7.03	7.14	7.24	7.34	7.43	7.51	7.59
7	6.30	6.43	6.55	6.66	6.76	6.85	6.94	7.02	7.10	7.17
8	6.05	6.18	6.29	6.39	6.48	6.57	6.65	6.73	6.80	6.87
9	5.87	5.98	6.09	6.19	6.28	6.36	6.44	6.51	6.58	6.64
10	5.72	5.83	5.93	6.03	6.11	6.19	6.27	6.34	6.40	6.47
11	5.61	5.71	5.81	5.90	5.98	6.06	6.13	6.20	6.27	6.33
12	5.51	5.61	5.71	5.80	5.88	5.95	6.02	6.09	6.15	6.21
13	5.43	5.53	5.63	5.71	5.79	5.86	5.93	5.99	6.05	6.11
14	5.36	5.46	5.55	5.64	5.71	5.79	5.85	5.91	5.97	6.03
15	5.31	5.40	5.49	5.57	5.65	5.72	5.78	5.85	5.90	5.96
16	5.26	5.35	5.44	5.52	5.59	5.66	5.73	5.79	5.84	5.90
17	5.21	5.31	5.39	5.47	5.54	5.61	5.67	5.73	5.79	5.84
18	5.17	5.27	5.35	5.43	5.50	5.57	5.63	5.69	5.74	5.79
19	5.14	5.23	5.31	5.39	5.46	5.53	5.59	5.65	5.70	5.75
20	5.11	5.20	5.28	5.36	5.43	5.49	5.55	5.61	5.66	5.71
24	5.01	5.10	5.18	5.25	5.32	5.38	5.44	5.49	5.55	5.59
30	4.92	5.00	5.08	5.15	5.21	5.27	5.33	5.38	5.43	5.47
40	4.82	4.90	4.98	5.04	5.11	5.16	5.22	5.27	5.31	5.36
60	4.73	4.81	4.88	4.94	5.00	5.06	5.11	5.15	5.20	5.24
120	4.64	4.71	4.78	4.84	4.90	4.95	5.00	5.04	5.09	5.13
∞	4.55	4.62	4.68	4.74	4.80	4.85	4.89	4.93	4.97	5.01

271

Table N

Critical Values for the Tukey Test   *Continued*

$\alpha = 0.10$

v \ k	2	3	4	5	6	7	8	9	10
1	8.93	13.44	16.36	18.49	20.15	21.51	22.64	23.62	24.48
2	4.13	5.73	6.77	7.54	8.14	8.63	9.05	9.41	9.72
3	3.33	4.47	5.20	5.74	6.16	6.51	6.81	7.06	7.29
4	3.01	3.98	4.59	5.03	5.39	5.68	5.93	6.14	6.33
5	2.85	3.72	4.26	4.66	4.98	5.24	5.46	5.65	5.82
6	2.75	3.56	4.07	4.44	4.73	4.97	5.17	5.34	5.50
7	2.68	3.45	3.93	4.28	4.55	4.78	4.97	5.14	5.28
8	2.63	3.37	3.83	4.17	4.43	4.65	4.83	4.99	5.13
9	2.59	3.32	3.76	4.08	4.34	4.54	4.72	4.87	5.01
10	2.56	3.27	3.70	4.02	4.26	4.47	4.64	4.78	4.91
11	2.54	3.23	3.66	3.96	4.20	4.40	4.57	4.71	4.84
12	2.52	3.20	3.62	3.92	4.16	4.35	4.51	4.65	4.78
13	2.50	3.18	3.59	3.88	4.12	4.30	4.46	4.60	4.72
14	2.49	3.16	3.56	3.85	4.08	4.27	4.42	4.56	4.68
15	2.48	3.14	3.54	3.83	4.05	4.23	4.39	4.52	4.64
16	2.47	3.12	3.52	3.80	4.03	4.21	4.36	4.49	4.61
17	2.46	3.11	3.50	3.78	4.00	4.18	4.33	4.46	4.58
18	2.45	3.10	3.49	3.77	3.98	4.16	4.31	4.44	4.55
19	2.45	3.09	3.47	3.75	3.97	4.14	4.29	4.42	4.53
20	2.44	3.08	3.46	3.74	3.95	4.12	4.27	4.40	4.51
24	2.42	3.05	3.42	3.69	3.90	4.07	4.21	4.34	4.44
30	2.40	3.02	3.39	3.65	3.85	4.02	4.16	4.28	4.38
40	2.38	2.99	3.35	3.60	3.80	3.96	4.10	4.21	4.32
60	2.36	2.96	3.31	3.56	3.75	3.91	4.04	4.16	4.25
120	2.34	2.93	3.28	3.52	3.71	3.86	3.99	4.10	4.19
∞	2.33	2.90	3.24	3.48	3.66	3.81	3.93	4.04	4.13

Table N

Critical Values for the Tukey Test   *Continued*

$\alpha = 0.10$

ν \ k	11	12	13	14	15	16	17	18	19	20
1	25.24	25.92	26.54	27.10	27.62	28.10	28.54	28.96	29.35	29.71
2	10.01	10.26	10.49	10.70	10.89	11.07	11.24	11.39	11.54	11.68
3	7.49	7.67	7.83	7.98	8.12	8.25	8.37	8.48	8.58	8.68
4	6.49	6.65	6.78	6.91	7.02	7.13	7.23	7.33	7.41	7.50
5	5.97	6.10	6.22	6.34	6.44	6.54	6.63	6.71	6.79	6.86
6	5.64	5.76	5.87	5.98	6.07	6.16	6.25	6.32	6.40	6.47
7	5.41	5.53	5.64	5.74	5.83	5.91	5.99	6.06	6.13	6.19
8	5.25	5.36	5.46	5.56	5.64	5.72	5.80	5.87	5.93	6.00
9	5.13	5.23	5.33	5.42	5.51	5.58	5.66	5.72	5.79	5.85
10	5.03	5.13	5.23	5.32	5.40	5.47	5.54	5.61	5.67	5.73
11	4.95	5.05	5.15	5.23	5.31	5.38	5.45	5.51	5.57	5.63
12	4.89	4.99	5.08	5.16	5.24	5.31	5.37	5.44	5.49	5.55
13	4.83	4.93	5.02	5.10	5.18	5.25	5.31	5.37	5.43	5.48
14	4.79	4.88	4.97	5.05	5.12	5.19	5.26	5.32	5.37	5.43
15	4.75	4.84	4.93	5.01	5.08	5.15	5.21	5.27	5.32	5.38
16	4.71	4.81	4.89	4.97	5.04	5.11	5.17	5.23	5.28	5.33
17	4.68	4.77	4.86	4.93	5.01	5.07	5.13	5.19	5.24	5.30
18	4.65	4.75	4.83	4.90	4.98	5.04	5.10	5.16	5.21	5.26
19	4.63	4.72	4.80	4.88	4.95	5.01	5.07	5.13	5.18	5.23
20	4.61	4.70	4.78	4.85	4.92	4.99	5.05	5.10	5.16	5.20
24	4.54	4.63	4.71	4.78	4.85	4.91	4.97	5.02	5.07	5.12
30	4.47	4.56	4.64	4.71	4.77	4.83	4.89	4.94	4.99	5.03
40	4.41	4.49	4.56	4.63	4.69	4.75	4.81	4.86	4.90	4.95
60	4.34	4.42	4.49	4.56	4.62	4.67	4.73	4.78	4.82	4.86
120	4.28	4.35	4.42	4.48	4.54	4.60	4.65	4.69	4.74	4.78
∞	4.21	4.28	4.35	4.41	4.47	4.52	4.57	4.61	4.65	4.69

Table O

Factors for Computing Control Limits

Number of Observations in Sample, n	$\overline{X}$ Chart Factors for Control Limits	R Chart Factors for Control Limits	
	A_2	D_3	D_4
2	1.880	0	3.267
3	1.023	0	2.575
4	0.729	0	2.282
5	0.577	0	2.115
6	0.483	0	2.004
7	0.419	0.07	1.924
8	0.373	0.36	1.864
9	0.337	0.184	1.816
10	0.308	0.223	1.777
11	0.285	0.256	1.744
12	0.266	0.284	1.716
13	0.249	0.308	1.692
14	0.235	0.329	1.671
15	0.223	0.348	1.652
16	0.212	0.364	1.636
17	0.203	0.379	1.621
18	0.194	0.392	1.608
19	0.187	0.404	1.596
20	0.180	0.414	1.586
21	0.173	0.425	1.575
22	0.167	0.434	1.566
23	0.162	0.443	1.557
24	0.157	0.452	1.548
25	0.153	0.459	1.541

Source: From Beyer, W. H., *Handbook of Tables for Probability and Statistics, 2nd Edition,* CRC Press, Boca Raton, Florida, 1986. With permission.